WITHDRAWN

D0184736

INSTITUTIONS IN ENVIRONMENTAL MANAGEMENT

Do environmental policy makers practise what they preach? What form should institutions supporting sustainable development take?

Institutions in Environmental Management explores the complexities of solving contemporary environmental problems within existing institutions, questions guidelines set out in recent influential policy reports, and suggests new agendas for sustainability, industrial ecology and institutional reform.

Including case studies from the USA, Europe and China, this book investigates a wide range of environmental problems presently confronting experts worldwide. Drawing on in-depth thematic interviews with environmental decision makers, the author analyses the problems with which these individuals are grappling and explains the mental models which they form to authorize and rationalize their decisions. Such mental models are seen to reinforce existing environmental institutions and policies, blocking beneficial institutional changes, limiting the range of policy options that experts perceive to be at their disposal, and constraining new agendas which might tackle long-term environmental problems. The book is organized into three sections: a theoretical section which introduces the basics of the new institutionalist theory and its relevance to environmental management; empirical case studies of modern environmental management; and a prescriptive section which draws out institutional reform principles.

Examining influential reports such as the Brundtland Commission's report and Agenda 21 set out at Rio, the author argues that long-term environmental concerns have been excluded from virtually all recent policy considerations and essential institutional reform is now needed to allow environmental experts and policy makers the autonomy to act upon what they believe to be sustainable environmental management.

Janne Hukkinen is Director of the Arctic Centre, University of Lapland.

ROUTLEDGE/EUROPEAN UNIVERSITY INSTITUTE STUDIES IN ENVIRONMENTAL POLICY

Edited by Jonathan S. Golub

Lecturer in Politics at Reading University and former Research Fellow at the European University Institute, Florence, Italy and the Max Planck Institute for the Study of Societies, Cologne, Germany.

1 DEREGULATION IN THE EUROPEAN UNION
Environmental perspectives
Edited by Ute Collier

2 GLOBAL COMPETITION AND EU ENVIRONMENTAL POLICY
Edited by Jonathan S. Golub

3 NEW INSTRUMENTS FOR ENVIRONMENTAL POLICY IN THE EU
Edited by Jonathan S. Golub

4 INSTITUTIONS IN ENVIRONMENTAL MANAGEMENT
Constructing mental models and sustainability
Janne Hukkinen

INSTITUTIONS IN ENVIRONMENTAL MANAGEMENT

Constructing Mental Models and Sustainability

Janne Hukkinen

London and New York

First published 1999
by Routledge
11 New Fetter Lane, London EC4P 4EE

Simultaneously published in the USA and Canada
by Routledge
29 West 35th Street, New York, NY 10001

Typeset in Garamond by J&L Composition Ltd, Filey, North Yorkshire
Printed and bound in Great Britain by
Biddles Ltd, Guildford and King's Lynn

British Library Cataloguing in Publication Data
A catalogue record for this book is available from the British Library

Library of Congress Cataloging in Publication Data
Hukkinen, Janne.
Institutions in environmental management: constructing mental models and sustainability / Janne Hukkinen.
p. cm.
Includes bibliographical references and index.
1. Environmental management. 2. Environmental policy.
3. Industrial ecology. I. Title.
GE300.H85 1998 98–39831
363.7′05—dc21

ISBN 0 415 16413 3

IN MEMORY OF MY BROTHER AND HERO,
KALLE (1956–97)

CONTENTS

FIGURES

TABLES

PREFACE AND ACKNOWLEDGEMENTS

This book is a synthesis of case studies I have conducted in the United States, Europe and China since the late 1980s on institutional issues of long-term environmental management and sustainable development. My aim is first of all to convince environmental policy analysts and professionals that, as we enter the twenty-first century and face a barrage of environmental problems that are longer in term, larger in scale, more complex and less well understood than before, the cognitive aspects of our collective efforts to understand and solve these problems matter more than ever. More to the point, the mental models that experts and decision makers utilize to understand environmental problems reinforce the existing institutional capacities to tackle the problems. But those same reinforced institutions in turn delimit the range of policy options the experts perceive to be at their disposal. Second, I wish to advise environmental policy analysts and professionals on how to investigate this feedback between individual mental models and their institutional boundary conditions. Third, similarities in the case study findings have persuaded me to outline institutional reform principles that I think would put us in a better position than we are today to face the challenge of sustainability. While the book is a monograph, I have tried to write it using language that is accessible enough to appeal to a multidisciplinary audience of academics and working professionals in government and enterprises. It can also serve as an advanced textbook for a graduate or post-graduate course on environmental institutions, policy or management.

To uncover the mental models of environmental experts, I listened to what they had to say about the environmental problems they were grappling with. Underlying their stories are the mental models with which they authorize and rationalize decisions. These models do not always stem from empirically verified reality, as many of the experts would like to claim, but rather from what they perceive to be that reality. The experts are processing incomplete information through mental models that can lead to choices that reinforce existing institutions as the only source of stability and certainty. And, from the point of view of most people, those institutions can be very inadequate and far from optimal. For these reasons, I do not treat individual stories as elements of a plausible theory of what the future might hold. Instead, I consider them as equal elements of a collective, and often internally inconsistent, model of and for the future. Analysis of the factual, attitudinal and logical inconsistencies of the collective model indicates how institutions constrain long-term policy decisions and what institutional changes might relax the constraints.

The main finding of this study is that institutions emphasizing short-term economic well-being persuade environmental decision makers to act against what they believe is sustainable environmental management. But in doing so the decision makers only support the same short-sighted institutions that constrain them in the first place. The recommendation that follows from this finding questions the guidelines of some of the most influential environmental policy reports of recent times, such as the Brundtland Commission's report and the Rio Conference's Agenda 21, both of which advise institutional integration of environmental and economic concerns. The case studies in *Institutions in Environmental Management* indicate that this may not be the solution to contemporary environmental problems. In fact, it may make the problems all the more intractable. Today's environmental and economic policy makers are already in intimate contact, but long-term environmental concerns, which are the backbone of sustainability, have been structurally excluded from virtually all policy consideration in the case study countries. Instead of putting sustainability on the agendas of sectoral agencies, sustainability concerns in those agencies should be identified and given collectively to a separate body endowed with a sustainability agenda. This would not only give prominence to sustainability and lay the groundwork for elaborating sustainable development in a democratic setting, but could also stimulate the genuine cross-sectoral communication that the Brundtland and Rio panels correctly identify as a prerequisite for resolving problems of environmental and economic development.

Institutions in Environmental Management will appeal to two academic audiences that have so far been rather far removed from one another. The first audience comprises researchers of comparative politics and international relations, who have recently put much effort into comparative studies of the effectiveness of international environmental regimes (Haas 1990; Haas *et al.* 1993; Young 1994). The second audience are environmental sociologists who have begun to ask how the new environmental regimes redefine and redistribute social power (Beck 1992; Beck *et al.* 1995; Dryzek 1997). In the first group, Haas (1990) argues that the existence of what he calls an 'epistemic community' promotes more effective collective responses to international environmental problems. In the case studies of *Institutions in Environmental Management*, I found similar patterns of cognitive response to long-term environmental problems – or 'epistemic communities' – in different cultures around the world, which led me to recommend the build-up of institutional support to such communities to facilitate long-term environmental decisions. But, as the students of comparative politics and international relations themselves point out, one should ask not only whether a regime is effective at solving the problem it was designed for, but whose agenda is being fulfilled or exluded and what impact the regime has on the distribution of power (Levy 1996). The case studies in this book illustrate the difficulties that analysts encounter in getting the institutional reform message across to decision makers who interpret the message as a threat to their power. I will recount examples of how a 'dialogical public sphere' (Beck *et al.* 1995) can get parts of the existing power structure to sympathize with reform.

The book itself is a story of my ten-year physical and intellectual journey across three continents, during which many individuals gave me a helping hand. The journey started in California in the late 1980s, when I was collecting data for my PhD dissertation at the University of California, Berkeley. What were meant to be preliminary interviews to acquaint myself with the state's agricultural drainage problems turned out to be the main data of my dissertation. I am indebted to Gene Rochlin and Emery Roe for convincing me the interviews were *the* data and for labouring with me to make sense of it all. In the process, they educated, challenged, and inspired me beyond measure, and forced me to look constantly for a better understanding of what the problem really was. Thanks are also due to David Jenkins, William Oswald and Matt Gerhardt, who helped me keep the engineering details straight in interdisciplinary assessment of environmental technologies, and to Michael Hanemann, who educated me on the socio-political complexities of the California case and helped me refine the arguments of the Colorado case study.

The journey continued on to Finland in 1991, when the Ministry of the Environment asked me to study the long-term future of Finnish waste management. I am not sure that I delivered what they had anticipated. What intrigued me was the idea of looking at individual scenarios in an unconventional way, namely as indicators of the capacity of today's institutions to facilitate long-term environmental policy. This also allowed me to extend the methods I had developed in collaboration with Emery Roe and Gene Rochlin from the regime of existing environmental problems to one of anticipated environmental problems. I thank Juha Koponen for initiating and encouraging the study and Christer Bengs and Jonathan Schiffer for showing a profound and critical interest in what I was doing and proposing.

A couple of years later I had the opportunity to put the Finnish study in a Europe-wide context during my stay in the Netherlands, and also to contrast the Western cases with a Chinese one when working on a Dutch development cooperation project in Hunan Province in 1994 and 1995. Raimo Lovio and Eva Heiskanen inspired me to look into green product concepts, Harald Sander provided critical macro-economic comments on my institutional analysis, and Pauline de Jong and Maarten Wolsink gave valuable insights on the corporatist characteristics of Dutch environmental management. I am grateful to them all. I thank Ma Lan and Eduard Vermeer for enlightening discussions that shaped the chapter on Chinese environmental management, and Guihong Chi for critical comments on my analysis. I am also thankful to the staff of the Sino-Dutch Management Training Project for valuable help in finding Chinese literature, arranging interviews, and providing translation and interpretation.

I am grateful to the following individuals, who read an earlier draft of the book with a critical eye and were a source of extremely helpful ideas on how to improve the manuscript: Chuck Dyke, Yrjö Haila, Susan Hanna, Risto Heiskala, Matti Kamppinen, Douglass C. North, Elinor Ostrom, Gene Rochlin, Emery Roe and Oran R. Young. I received helpful editorial comments and support throughout the writing from Sarah Lloyd and her assistants at Routledge, for which I thank them. I am also deeply indebted to the environmental experts and policy makers in California, Colorado, Finland and China for the time and consideration they were willing to devote to the long interview sessions. In addition, there have been many others who have contributed to the effort along the way. I thank them all collectively.

Some chapters of this book are based on substantially revised excerpts from earlier articles of mine published by international peer-reviewed journals of environmental policy, economics, science and technology. I thank the American Society of Civil Engineers (ASCE) for permission to publish parts of 'Institutional distortion of drainage modeling in Arkansas River Basin', *Journal of Irrigation and Drainage Engineering*, Vol. 119, No. 5 (1993), pp. 743–55 in Chapter 4 and parts of 'Sociotechnical analysis of irrigation drainage in Central California', *Journal of Water Resources Planning and Management*, Vol. 117, No. 2 (1991), pp. 217–34 in Chapter 5; Academic Press Limited for permission to publish parts of 'Bayesian analysis of agricultural drainage problems in California's San Joaquin valley', *Journal of Environmental Management*, Vol. 37 (1993), pp. 183–200 in Chapter 5; Elsevier Science B.V. for permission to publish parts of 'Corporatism as an impediment to ecological sustenance: the case of Finnish waste management', *Ecological Economics*, Vol. 15, No. 1 (1995), pp. 59–75 in Chapter 6; the Soil and Water Conservation Society for permission to publish parts of 'Irrigation-induced water quality problems: can present agencies cope?', *Journal of Soil and Water Conservation*, Vol. 46, No. 4 (1991), pp. 276–8 in Chapter 8; and John Wiley & Sons, Ltd for permission to publish parts of 'Green virus: Exploring the environmental product concept', *Business Strategy and the Environment*, Vol. 4, No. 3 (1995), pp. 135–44 in Chapter 9, and parts of 'Long-term environmental policy under corporatist institutions', *European Environment*, Vol. 5, No. 4 (1995), pp. 98–105 in Chapter 10. I also thank US Geological Survey for permission to reprint Figure 1 of A.W. Burns, *Calibration and Use of an Interactive–Accounting Model to Simulate Dissolved*

Solids, Streamflow, and Water-Supply Operations in the Arkansas River Basin, Colorado, US Geological Survey Water-Resources Investigations Report 88–4214, Lakewood, CO: US Geological Survey (1989) as Figure 4.1 of this book, and the American Society of Civil Engineers (ASCE) for permission to reprint Figure 1 of S.B. Moore, 'Selenium in agricultural drainage: essential nutrient or toxic threat?', *Journal of Irrigation and Drainage Engineering*, Vol. 115, No. 1 (1989), pp. 21–8 as Figure 5.1 of this book.

I am grateful to the following organizations for their financial support during the completion of the case studies and the book: Maj and Tor Nessling Foundation, Academy of Finland, Finnish Cultural Fund, University of California Berkeley Foreign Student and Scholar Services, California Department of Water Resources, US Environmental Protection Agency, Technology Development Centre of Finland, Ministry of the Environment of Finland, and Dutch International Development Cooperation Agency.

My ten-year struggle with the research and writing of this book coincides but pales in comparison with my brother Kalle's mortal battle with a brain tumour. These pages were written in his memory, with loving respect.

Last but certainly not least, my love and gratitude go to Tuula, Meri and Tuomas, who did not seem to mind too much the dinners we had together when my involvement in the writing of this book was at its height, when my body was present but mind was not. Even then, I felt their love and support.

Rovaniemi, Finland
January 1998

1

INTRODUCTION

When my family and I moved to the town of Rovaniemi in northern Finland, which the locals claim to be where Santa Claus lives and rules, my five-year-old daughter Meri announced she did not like Santa. In her mind, he threatens children by telling them they do not get presents unless they are nice. And one is not supposed to threaten people. Since Christmas was just around the corner, I told her she should discuss this with Santa. But when Santa came on Christmas eve, Meri sat silently and respectfully on her aunt's lap. When I asked her later why she had not brought up the issue, she said she did not want to – but made it clear she still did not like Santa's threats.

I was amazed that a five-year-old can see the moral inconsistency many educators and other authorities are guilty of, when they violate the rules they preach. I had just begun writing this book and my mind was replaying the interviews I had conducted over the past decade – and here was a child behaving the way the adult interviewees in government agencies and private enterprises in the US, Europe and China had behaved. When an individual is in a situation where the strength of institutional rules is tangible, such as a child facing Santa on Christmas eve or a bureaucrat facing the political sensitivities of an environmental management decision, the person tends to go along with the rules and feels comfortable doing so. But when in a situation where institutional constraints are remote, such as a daughter chatting with father or a bureaucrat with researcher, the individual feels free to reflect critically on the rules.

This book is about the ways in which institutional constraints influence how individuals think and act in environmental organizations, and how their thinking and acting in turn reinforce the same institutional constraints. Environmental institutions are the rules that determine which organizations in the society can have a say in environmental management decisions and what types of environmental management are considered legitimate. The mental models of environmental decision makers and the formal institutions within which they make decisions are in a mutually reinforcing feedback. It is a feedback no different from the one Santa Claus is locked in: the more their children believe in Santa, the more reason parents have to maintain the practices that support the institution of Santa; and the more ingrained the rituals around Santa, the firmer the children's faith.

But as parents with growing children can testify, the reinforcing feedback between institutions and thinking is not one that guarantees institutional stagnation. Given an appropriate situation, individuals will voice their reservations about the limits that institutions impose on their actions, the way my daughter did. Institutions are resilient, because individuals tend to follow them; but this does not mean individuals could not critically reflect upon themselves and their relationship with institutions. Human capacity for self-reflection is the seed of institutional change.

Self-reflecting individuals in environmental organizations are the methodological focus of this book. Since modern environmental problems are complex and uncertain, decision makers often find that not even the best scientists can provide the facts to support decisions – or when they can, the facts obtained from different experts are conflicting (Douglas and Wildavsky 1983; Haila and Levins 1992; Norgaard 1994; Redclift 1992). As a result, the only solid bases policy makers can construct for their decisions are the mental models and stories they and their colleagues use to articulate and make sense of the complexity and uncertainty (Roe 1994). The mental models of professional elites, i.e. government officials, enterprise managers, researchers, consultants and representatives of non-profit organizations, shape and are themselves shaped by the institutional rules of environmental management. Depending on their disciplinary inclination, analysts have used different terms to describe this interactive relationship. Sociologists such as Berger and Luckmann (1967) call it a dialectical relationship between human cognition (the producer) and the institutional world (the product), in which the product acts back upon the producer. Institutional economists such as North (1992) identify it as a feedback between mental models and institutions. While the feedback has been acknowledged as one of the central ingredients of institutional analysis, few studies have addressed the issue explicitly and in depth. The case studies in *Institutions in Environmental Management* all rely on methodologies that probe the relationship between individual cognitive models and formal institutions.

Since the late 1980s I have conducted case studies on the institutional aspects of environmental policy and management in the US, Europe and China. There are, of course, significant differences in culture, institutional setting and the level of social and economic development, particularly between the Western cases and the Chinese case. Furthermore, the studies cover substantially very different issues, ranging from agricultural to industrial environmental management, and from government to corporate environmental policy. Despite these differences, the cases share the properties of today's most pressing environmental challenges: the problems are created by complex and poorly understood ecological, technical, social and psychological relationships extending over large areas and far into the future. According to the case studies, the institutional reforms required to manage environmental issues of this kind run against the accepted wisdom of many recent panels on global environmental policy, such as the Brundtland Commission and the Rio Conference (Robinson 1993; World Commission on Environment and Development 1990). Where these panels recommend the institutional integration of environmental and economic concerns, this book argues for a clear separation of the two and the establishment of an autonomous agent with the sole responsibility to pursue long-term environmental policy. Such autonomy would elicit honest professional opinions from the members of the long-term environmental decision making body.

Institutions in Environmental Management is a novel contribution to the study of the feedback between mental constructs and environmental institutions, and to the design of institutions of contemporary environmental management. The book provides practical advice on how to understand and analyse the interaction between institutional and cognitive processes in environmental management and how to draw policy recommendations from the analysis. To that end, the book's chapters are organized in three parts: theoretical chapters introducing the basic tenets of the new institutionalist theory and their relevance to environmental management; empirical case studies of modern environmental management; and prescriptive chapters drawing out institutional reform principles. Allow me to introduce each briefly.

PART I: INSTITUTIONS AND THEIR ANALYSIS

The literature on institutions indicates that there is a lack of empirically based understanding of the mental feedback process as it weaves in and out of the institutional setting (Berger and Luckmann 1967; North 1992; Rutherford 1996). *Institutions in Environmental Management* provides that basic understanding. The book's cases show that an already existing theory – primarily North's (1992) articulation of institutions and institutional change – can be applied to explain institutional phenomena in environmental management and to gain insights on institutional reform.

Institutions in Environmental Management bridges new institutionalist theory and methodology. The new institutionalists understand society as a game and institutions as the rules that guide the game (North 1992; Ostrom 1994). While rules can be further categorized into several classes, there is an underlying logical structure unifying all rules. That structure is causality, which in its simplest form has the structure 'If A, then B'. Cognitive psychologists have found that whenever individuals describe their field of expertise in narrative form, a causal structure underlies the narratives (Bower and Morrow 1990). The methods of cognitive mapping I have used in the case studies reveal the cause-and-effect rules with which decision makers understand decision making problems and on which they base their policy decisions. In this way the rules teased out of the experts' narratives are individual interpretations of the institutions within which they make decisions. Analysis of the mental models reveals not only the meaning of institutional rules to key decision makers, but also the impact the rules have on their decisions.

Putting mental models in the context of formal and informal institutional rules illustrates how cognitive mapping points directly toward institutional reform. Formal institutions are the laws, regulations and other codified rules of the society. Informal rules are unwritten customs and administrative procedures. The less correspondence between the formal and informal institutions, the more likely social tensions and pressures toward institutional reform (North 1992). Since informal rules are not codified, they exist only in the minds and actions of individuals. Investigating what people say provides a vantage point on the informal rules that individuals can consciously describe, whereas studying what people do reveals the routines that structure the everyday lives of individuals but that individuals themselves cannot express in words (Giddens 1986).

In the case studies of *Institutions in Environmental Management* the difference between what people say and do is less clear. After all, the cases are based on interviews with public and corporate administrators who spend most of their professional lives in the meetings of committees, task forces, office departments, public hearings and so forth, where their main activity is speaking, thinking and making decisions. The closest one can come to codifying the informal rules of such experts is explicit analysis of the mental models contained in individual minds. This is what cognitive mapping is all about. The cause-and-effect networks drawn from the individual narratives reflect the informal rules that guide decision making in a particular environmental management regime. Analysing the logical inconsistencies and the cognitive dissonances in the mental models and comparing them with the formal institutional rules pinpoints very directly where institutional tensions exist in that particular decision making regime and how those tensions might be resolved through institutional reforms.

The principal institutional proposition of *Institutions in Environmental Management* centres around the incompatibility of formal and informal institutions. The four case studies presented in the book indicate short-term economic profit principles tend to dominate the formal environmental institutions. At the same time, many environmental decision makers and experts hold professional convictions about the principles of long-term environmental sustainability, which are in sharp contrast with the short-term economic principles. The professionals have developed an elaborate informal rule to deal with the pull between their own long-term convictions and the short-term pressure from the formal institutions. They rationalize their decisions with what are cognitively dissonant mental models. On the one hand, they hold on to their professional convictions about environmental sustainability; on the other hand, the decisions they make yield to the short-term economic principles imposed by the formal institutions. Accordingly, cognitive dissonance arises as a reaction to the high professional and organizational sacrifice the individual would have to make for acting upon his or her professional convictions. But the ability to maintain this cognitive dissonance also ends up reinforcing the same formal institutions that triggered it: day-to-day decisions made in accordance with short-term economic principles are the threads reinforcing the fabric of formal institutions that, in turn, enable cognitively dissonant decision makers to continue to make decisions.

Part I contains two chapters. Chapter 2 ('Institutional change and expert thinking') is an introduction to contemporary theory of institutions and institutional change, particularly its relevance to environmental policy and management. The chapter describes what existing literature tells us about the institutions of modern environmental management, how they can be analysed, and how they change or can be changed. Particular attention is paid to the feedback between the mental models of individual decision makers and the institutions within which they operate. This leads to a summary of the book's theoretical framework for analysing the cases: environmental management is understood as the set of 'means–ends programmes' designed to remedy perceived environmental problems. Environmental institutions refer to the rules that guide the design of these means–ends programmes. Environmental institutions and environmental management choices made by decision makers are in a feedback, where management choices influence institutions, and *vice versa*. Chapter 3 ('Finding the institutional rules') describes the methodologies used in the case studies to analyse the feedback, including cognitive mapping and network analysis. Cognitive mapping takes the stories of individual decision makers as the data of the analytic exercise. Each individual story is based on a mental model with which that particular decision maker rationalizes his or her environmental management decisions. The mental models compose interrelated issues or problems that can for analytical purposes be coded as either issue networks or causal networks. Analysis of the inconsistencies and dissonances of the mental models indicates how institutions constrain long-term environmental policies and how the constraints might be relaxed.

PART II: CASE STUDIES ON MODERN ENVIRONMENTAL MANAGEMENT

The case studies of *Institutions in Environmental Management* demonstrate the modalities of the mental–institutional feedback, the pragmatic ways of analysing the feedback, and the reasoning behind policy recommendations drawn from the analysis. Although the cases come from three different continents – North America, Europe and Asia – they do not constitute a comparative cross-national study of environmental institutions and management. The purpose of the diverse material is rather to compare and contrast environmental institutions within the same methodological and theoretical framework, namely, cognitive mapping and institutional theory. And while the case studies do not guarantee ways of bringing about institutional change, they do lead the way to the institutional reform principles of Part III and indicate potential pitfalls in putting the reforms in action.

The four case studies conducted from the mid-1980s to the mid-1990s represent diverse environmental issues and locations, including agricultural toxics management in California, water quality modelling in Colorado, the development of long-term waste management strategies in Finland, and environmental protection in China. The unifying characteristics of the cases are complexity, uncertainty and relatively large spatial and temporal scales. The social and biophysical relationships that characterize the environmental issues are complicated and poorly understood – not just in terms of empirically verifiable biophysical relationships, but also in terms of relationships that the actors involved in environmental management believe to be underlying the issues. The spatial areas in the case studies are large and contain diverse ecosystems. Finland's hazardous waste management system, for example, relies on a centralized plant that treats much of the hazardous wastes of this 338,000 square kilometer nation at a single location in southern Finland. Irrigated farmland in California's Central Valley covers just 19,000 square kilometers, but the complex water conveyance system transmits the impact of irrigation and drainage throughout most of the 411,000 square kilometers of the state and, in fact, beyond its borders to most of the arid western US. Finally, all the cases deal with long temporal scales, simply because the implications of the environmental issues span generations. The claim for intergenerational impact is here qualitatively different from the platitude that all environmental problems have long-term implications. Experts interviewed in the book's case studies insist that any *operational* solution to the environmental problems needs to involve consideration of the very long-term future. The intergenerational aspect constitutes a decision making problem significantly different from so many benefit–risk comparisons concerning a single generation.

The case descriptions underscore how the book's basic institutional argument is realized through very different sociopolitical contexts around the world. Environmental decision makers and experts in California, Colorado, Finland and China adhere to cognitively dissonant mental models to rationalize their decisions. In some cases, cognitive dissonance is evident in internally contradictory statements about the central decision issues. A Californian irrigation official, for example, argued that it was impossible to construct a system-wide waste water channel to take polluted drainage out of California's Central Valley into the Pacific Ocean, although every task force over four decades had found that such a channel would be the only way to manage the valley's agricultural drainage effectively over the long term. At the same time, the official argued that it was none the less imperative to manage the valley's agricultural drainage effectively over the long term. In other cases, individual experts described the issues in circular arguments that were held together by fundamentally conflicting goals. A Finnish waste management expert, for example, described as a serious long-term problem the fact that natural resources are becoming ever scarcer in today's economy. This, according to the interviewee, creates an urgent need for sustainable waste management policy. Such policy, however, would trigger other problems by reducing the material standard of living and by running against efforts to revitalize the stagnant economy in the short run. In the expert's experience, economic revitalization must win in decision making, which only intensifies the depletion of natural resources.

However puzzling or even irrational these lines of thinking may appear, they are quite rational when considered in the context of formal institutions. All four cases show that, had the experts not acted upon the short-term operating assumptions despite their long-term convictions, they would have threatened not only their own professional credibility and standing, but also the legitimacy and survival of the organizations they represented.

Chapter 4 ('Institutional distortion of water quality modelling in southern Colorado') describes large-scale agricultural water and salt management in the Arkansas River Basin of southern Colorado. In methodological terms the case presents an elementary, qualitative approach to cognitive mapping, based on issue network analysis. In terms of the main argument of the book, the chapter highlights the incompatibility of sustainability criteria developed for local short-term problems on the one hand and large-scale long-term ones on the other. The book's second case study (Chapter 5: 'Network analysis of the controversy over irrigation-induced salinity and toxicity in central California') describes the management of toxic substances found in irrigation drainwater in central California. Here the methods are more sophisticated. A combination of qualitative narrative and quantitative Bayesian network analysis reveals the mental constructs of key decision makers in the drainage controversy. The chapter illustrates how organizational control of different stages of environmental technology can significantly influence the need for institutional change.

Chapter 6 ('Corporatism as an impediment to sustainable waste management in Finland') presents the third case, which focuses on Finnish decision makers' efforts to develop long-term solid and hazardous waste management strategies. The case shows how institutional change pressures can emerge from the administrative arrangements of environmental regulation and implementation. In addition, the chapter provides further empirical evidence for the relevance of the earlier-mentioned institutional change pressures. Finally, Chapter 7 ('Environmental management in China') discusses environmental management in China as an example of a single nation state in which serious large-scale environmental problems coincide with culturally ingrained corporatist institutions incapable of dealing with sustainability issues.

PART III: INSTITUTIONAL REFORM PRINCIPLES

The case studies show that the proximity of environmental decision makers to economic planners facilitates short-term environmental management and regulation, but hinders far-sighted environmental policy. The truly long-term environmental concern, which is the essence of sustainable development and an issue quite different from day-to-day regulatory compliance, needs to be institutionalized. What is needed, I argue, is the establishment of an autonomous social agency with the sole responsibility to pursue long-term environmental objectives.

The idea of creating autonomous social units as a way of facilitating long-term environmental policies is based on experiences obtained from institutional arrangements in regimes other than environmental policy. Analysts of the long-term performance of economic systems have observed an inverse relationship between the willingness of decision makers to reveal their convictions and the price formal institutions impose on decision makers who reveal their convictions: the higher the price individuals have to pay for revealing their convictions, the lower their tendency to do so; and, conversely, the lower the price, the higher the likelihood that professionals will reveal and act upon their convictions (North 1992). 'Price' here refers to an institutional cost, which for an individual can mean losing position, authority or professional prestige. Institutional analysts also point out that the formal institutional price can be lowered for individuals by providing them with institutional autonomy to reveal their convictions. This is why judges in many instances can make unpopular decisions, university professors teach what they think is important, central bankers set interest rates freely and voters cast their votes as they please.

Obviously, there is no single way of establishing autonomous environmental agents. The cases indicate that administrative redesign is adequate, when conflicts of interest in environmental management can be traced along administrative lines of duty. But administrative reforms also raise the spectre of conflict between technocratic rationality and democratic balancing of interests. What guarantees that an autonomous environmental management agency will not develop into a monolithic power base that many voters are sure to regard as inherently undemocratic and elitist? To counter such criticism, participatory systems of environmental policy and management are proposed that enable non-expert stakeholders in an environmental issue to influence decisions. When developing such systems, particular attention should be paid to creating incentive structures that allow individuals to optimize at the short-term margins yet prevent the sum and sequence of individual actions from exhausting the integrity and resilience of the natural resource base. Land use planning and environmental impact assessment are examples of existing procedures that could be developed into institutionalized forms of interaction between diverse interest groups in environmental management. This would obviously require a shift from today's project-based assessment to a permanent assessment body with elected members, entailing careful consideration of mandates between the new body and other elected bodies. The establishment of economic instruments is another way of exposing environmental policy to open parliamentary decision making.

Institutions in Environmental Management agrees with those who argue that no scientifically stable objective information can be obtained about long-term environmental policy due to its inherently political and dynamic nature (Norgaard 1994; Redclift 1992). The proposed institutional arrangement could, however, elicit honest professional beliefs from the decision makers. Although decision makers would still lack the solid backing of objective science, they would be in a much more informed position than they are today. As the case studies show, today's short-term economic imperatives put considerable institutional pressure on decision makers to compromise their convictions about the need to develop long-term environmental policies.

Living with the pull between short- and long-term pressures is, of course, nothing new to decision makers. But in the case of modern environmental management, relieving that tension is of the utmost importance. Institutionalizing sustainability would first of all enable environmental decision makers to take action on long-term environmental issues, something the case studies repeatedly show they cannot do as long as they remain in a state of cognitive dissonance. Second, institutional autonomy would facilitate open exchange of information – a key justification for having institutions in the first place – in the debates over sustainability. In contrast to many contemporary studies on environmental management that emphasize conflict resolution, this book argues that the psychological internalization of conflict as cognitive dissonance and the consequent absence of open political conflict secure the ambiguity of sustainable development and impede action on long-term environmental policy.

The book is not a blueprint for institutional design. It does, however, provide compelling evidence as to why existing institutions cannot achieve long-term environmental management objectives and what institutional design principles follow from the current failures. The point of the feedback between institutions and mental models is that reflection itself is affected by existing institutional constraints. But the institutions that arise out of the reflection are not necessarily economically or socially desirable. The case studies I present show time and again that decision makers may have very clear preferences about how environmental policy *should* develop, but institutional constraints persuade them to hold operating assumptions about how environmental policy *will* develop that are completely contradictory to their preferences. To reiterate, the tension between preferences and operating assumptions allows individual decision makers to develop quite rational short-term survival tactics for the organizations in which they operate while closing out options for long-term policy.

The reinforcing feedback between institutions and mental models also makes institutional change an inherently slow process. The case studies, for example, come from no longer than a decade ago, too short a time to provide empirical evidence of actual institutional change. None the less, the case studies indicate potential pitfalls in initiating such reforms. Both the California and Finnish case studies hit unexpected snags in progressing to the implementation stage, and detailed investigation of the processes that stalled implementation offers valuable guidance for institutional reform work. The US and European case studies also provide practical insights on how to bring about institutional change by transforming existing organizational structures and administrative procedures.

In Chapter 8 ('Principles of institutional reform'), institutional reform recommendations are developed on the basis of institutional theory, empirical experiences in areas other than environmental management, and the case studies developed in earlier chapters. The creation of institutionally independent social agents to act on behalf of long-term environmental concerns is argued for, and both organizational and economic reform ideas are developed. Chapter 9 ('Institutions of industrial ecology') synthesizes the case analyses into institutional design principles for contemporary environmental management. The design principles are presented in the conceptual framework of industrial ecology, to date treated primarily as a technological challenge. The industrial ecology viewpoint is an appropriate one for *Institutions in Environmental Management*, as it entails discussion of spatial and temporal scales with respect to institutional design. The chapter discusses the compatibility of institutional design principles for modern large-scale environmental management with those derived in earlier studies for traditional smaller-scale management, and presents a tentative synthesis of design principles for the two management regimes. But, as the case studies illustrate, the feedback between institutions and mental models provides good reason for decision makers to resist change in existing institutions, even if they believe those institutions cannot bring forth desirable environmental policy outcomes.

Chapter 10 ('Experts in public') investigates in detail the failures in institutional reform and their implications for the role of experts in public debates over environmental institutions. The chapter illustrates the limitations of institutional reform by focusing on long-term environmental policy and management in the European Union (with particular reference to the case of Finnish waste management) and the United States (with specific reference to the case of Californian agricultural drainage management). The EU case is a warning to those optimists who expect far-sighted European environmental policy to result from the replication of national environmental institutions at the transnational level. In the US case, agency reactions to the policy recommendations of the California drainage study are recounted at different levels of government bureaucracy. In the conclusion to Chapter 10, I argue that experts can none the less have a positive role by initiating public dialogue, not only over environmental institutions, but more generally over the rights, responsibilities, and power relationships that bear heavily upon how the environment is and can be managed.

Reminded by what books that rely on scattered notes and cross-references have put my fingers through, I have tried to compose the book in a way that minimizes the need for the reader to flip between pages. I do not refer to notes. I have included two Appendices, one with a description of the transformations that took place in the coding of narrative transcripts into problem networks, another with a detailed account of the formalism for presenting aggregated problem networks as Bayesian networks. These, however, are details for those interested in details. Fluent reading of the book requires no reference to them.

Part I

INSTITUTIONS AND THEIR ANALYSIS

2

INSTITUTIONAL CHANGE AND EXPERT THINKING

Institutions and sustainability are interwoven. Since the Brundtland Commission promoted the concept of sustainable development in 1987, it has been defined and redefined in numerous and often sharply conflicting ways (World Commission on Environment and Development 1990). All definitions, however, share notions of permanence, resilience and endurance over long time-scales. These are also the characteristics that connect sustainable development with institutions. Institutions, or the rules that define what are legitimate actions for organizations and individuals to take in society, often prove to be lasting and difficult to change (Berger and Luckmann 1967; North 1992). Therein lies their promise – and challenge – for those wishing to institute sustainable development by design: the task of changing today's institutions will be difficult, but once successfully and appropriately changed, the new or revised rules are likely to have an impact over centuries.

The conception of society as a game and institutions as the rules that constrain the actions of individuals and organizations in the game immediately raises issues of cognitive processes and mental models. After all, it is in the minds of individual decision makers that institutional rules have their most intimate influence on the actions individuals take in the society. Individuals understand institutional constraints by constructing mental models of the way the world around them works, and make decisions on the basis of these models. To a neoclassical economist, the outcome of the social game is determined by player preferences and market forces. The neoclassical model, however, assumes transparent information and honestly revealed preferences. Historical evidence frequently suggests otherwise. Individual economic actors typically lack the knowledge that is necessary to play the game and cannot express their preferences either fully or accurately (Peterson 1973). Institutions determine how individual economic actors can express their own views, which may result in very different outcomes to the game from those predicted by neoclassical economics and public choice theory alone (North 1992).

The case studies in *Institutions in Environmental Management* penetrate the moment when the institutional rules that guide experts materialize as environmental management decisions. The purpose of this chapter is to outline the processes by which environmental institutions influence expert thinking, and the processes by which the mental models of experts in turn reinforce those environmental institutions. The feedback between mental models and institutions has been recognized by institutional analysts as one of the central elements of institutional stability and change (Berger and Luckmann 1967; North 1992; Scott 1987). What has been lacking is an empirically based explanation of the feedback mechanism.

The case studies in this book show cognitive dissonance to be a central element in the feedback between expert thinking and institutions. The interviewees would like to act upon what they believe to be the principles of sustainability, but feel institutionally compelled to decide on the basis of short-term economic priorities. The stronger the experts perceive the institutional pressure to achieve short-term economic benefits to be, the more pronounced the cognitive dissonance they experience; but the more pronounced the cognitive dissonance, the more the experts think they must reduce the dissonance by obeying the prevailing institutional rules. This is how a Finnish waste management expert articulated the role of cognitive dissonance in contemporary environmental debates:

> So then we come down by surprise, and ask ourselves, 'Why did this happen? Why did that Chernobyl break down like that, when it was only supposed to generate electricity? We didn't build it for any other purpose!' And then we act as if we were surprised, saying, what the devil, now something else happened than what was intended. We can in a way blind ourselves and not care about it, although many people and experts say that we're all the time producing bad results as well.
> (interview 2)

I do not intend to present a theoretical overview of institutions and mental models in the following sections. I will focus strictly on explaining the feedback between mental models and environmental institutions that I observed in the case studies. Theoretical reference points only support the proposition of the feedback between cognition and environmental institutions and anchor the discussion to current institutional thinking. I will provide further reference points to institutional and organizational literature in later chapters when presenting the methodology of cognitive mapping and the case studies themselves.

TERMINOLOGY

I understand the term 'institutions' here the way that institutional theorists have defined it since the early 1900s. Institutions are the rules, i.e. predefined patterns of conduct, that the members of a social group have generally accepted (Berger and Luckmann 1967; Rutherford 1996). The rules can be informal rules, such as norms, habits and customs; or formal rules, such as written laws, regulations and standards. The rules apply in a game, which is society. Members of a social group or organization are like players in a team. Society is seen as a game played by organizations according to the rules laid out in institutions. Organizations devise strategies to win or survive in the game (North 1992).

Environmental institutions are social institutions operating within the regime of environmental policy and management (Young 1982). Environmental institutions of the formal kind are environmental statutes, regulations, performance standards and various formal administrative guidelines. At the most general level, informal environmental institutions are the customary ways of conceptualizing environmental policy and management. Such notions of 'conventional environmental management' unfold in the accepted ways in which we conceptualize, for example, the relationship between economic and environmental issues in policy debates, the allocation of authority among environmental agencies, or the organization of technologies required to achieve environmental management objectives. Environmental organizations that play the strategic games described in the case studies include regulatory agencies at national, regional and local levels of government, semi-governmental and governmental environmental management agencies, private corporations developing environmentally sound processes or products, environmental activist groups, research institutes and consultancies. To win or survive in the game, all of them develop environmental management strategies, variously called policies, programmes, plans, guidelines, strategies, and so on.

I must emphasize the broad conception of environmental management adopted in this study. I understand environmental management as the set of means and ends that environmental organizations make use of in their activities. Depending on what the key environmental issues are perceived to be, these activities focus on the structure and functions of the organization itself, the operations of the organization or the impact it has on its ecological and social environment (Table 2.1). The case studies in *Institutions in Environmental Management* revolve around each of these focal points. In the Californian case study, for example, environmental experts focused on the organization by developing strategies that would ensure the survival of the state's irrigation bureaucracy. But they also concentrated on the operations of that bureaucracy by designing better on-farm management practices that would reduce the accumulation of harmful salts in the soil. And they conducted numerous social and environmental impact studies to describe the diverse costs and benefits of irrigation and drainage management in California.

My understanding of environmental management thus incorporates two

Table 2.1 Environmental management as the means and ends by which an organization transforms itself, its operations and its environment

	Focus of activity		
	Organization	*Operations*	*Environment*
Objective	To improve the structure and functions of the organization as an actor in the environmental field	To improve environmental activities and performance of the organization	To reduce negative social and environmental impacts of the organization's activities
Examples	Environmental strategy Standards for environmental management systems Environmental audit Environmental accounting	Choice of environmental technology Environmentally sound product and process design Environmental monitoring and evaluation	Environmental impact assessment Social impact assessment Life-cycle assessment Benefit/cost analysis

different notions that appear in scholarly writing, one viewing environmental management as the caretaking of natural resources such as wetlands, grazing areas and forests (and exemplified by articles in such publications as the *Journal of Environmental Management*), the other regarding environmental management as the activities a firm undertakes to achieve environmentally sound production and products (e.g. the journal *Business Strategy and the Environment*). The case studies in this book span these poles, from environmental management as natural resource management to corporate environmental management.

Besides the division into formal versus informal, institutions can be categorized hierarchically into constitutional, collective choice and operational rules (Ostrom 1994). All these are found in the case studies and play a central role in the formulation of institutional design recommendations. Operational rules directly affect the day-to-day decisions over the appropriation of environmental resources and the regulation of environmental management – to the extent that it is often difficult to distinguish an operational-level environmental institution from environmental management, as in the case of the corporate environmental management standard ISO 14001, which specifies the design and operation of a firm's environmental management system. Collective choice rules indirectly influence

Table 2.2 Classification of environmental institutions as the rules that guide individuals and organizations in environmental management, with examples

	Formal: written laws, regulations and standards	*Informal: unwritten norms, habits and customs*
Constitutive (What is the environmental policy agenda and who is eligible to participate in resource use?)	AGENDA 21	ENVIRONMENTAL MORALS
Collective choice (What is the environmental policy and what are the design principles of the environmental management system?)	ENVIRONMENTAL POLICY OF A FIRM	SUPPORT TO AN ENVIRONMENTAL INTEREST GROUP
Operational (What are the operational details of the utilization of environmental resources?)	ENVIRONMENTAL MANAGEMENT STANDARD (SUCH AS ISO 14001)	SOURCE SEPARATION OF HOUSEHOLD WASTE

operational rules. Environmental managers and authorities use collective choice rules to design environmental policies and management systems. Finally, constitutional rules affect the design of collective choice rules, and determine who is eligible to participate in environmental management and how such issues are formulated. Institutional rules are organized in a nested hierarchy (Table 2.2). Changes at one level of rules are always influenced by another set of rules at a deeper level. A firm's decision to adopt the ISO 14001 standard, for example, most likely reflects management-level commitment to a corporate environmental policy. And significant numbers of people would be unlikely to support environmental interest groups without adhering to a set of ethical principles in relation to the environment.

Such categories of institutions may give the impression of a stable system of rules. This is not the case. Institutions do change, albeit slowly, and categorizations such as ours help explain the dynamics that economic and social historians have observed in institutional change, our next topic.

INSTITUTIONAL CHANGE

This book concentrates on institutional change, because the initiatives it takes as its subject were originally launched in response to practical environmental policy problems calling for solutions. The orientation on perceived environmental problems makes *Institutions in Environmental Management* clearly prescriptive: its methodological and analytical approaches were developed in the belief that institutional change is not just evolutionary, but is also influenced by human intentions. However, since institutional change, of both the planned and the evolutionary kind, is necessarily slow, it is impossible to present empirical evidence of the successes or failures of the proposed institutional recommendations. As will become clear in Chapter 10, only the first steps on the path of institutional redesign can be recorded, namely agency reactions to the proposed changes.

What drives institutional change, for our purposes, is the tension between formal and informal institutions (North 1992). When formal rules are inconsistent with the informal constraints that guide individual behaviour, the unresolved tension between them leads to long-term political instability. Resolving the tension will always be slow, because all institutions resist change. Informal institutions, however, are more resilient than the formal ones. Laws and regulations change and can be changed more easily than the way people have grown accustomed to think and act in their everyday lives.

To appreciate how institutional change comes about, we need to understand the informal constraints and their relationship with the formal ones. Since informal institutions are not written down, the only place they are recorded is in individuals' minds. The closest an analyst can come to the informal rules is by deciphering them from the mental constructs the individuals reveal in their stories. As the case studies illustrate, analysis of the mental models with which individual decision makers and experts make sense of environmental problems also reveals the informal constraints to their thinking. What is even more important in terms of assessing the potential for institutional change is that such an analysis pinpoints the flashpoints between informal and formal rules. One of the most problematic features of this tension is that it creates incentives for individuals to hide information about their honest beliefs and preferences.

Historical analyses of the influence of institutions on economic performance indicate that there is an inverse relationship between the willingness of individuals to follow their convictions and the price formal institutional constraints impose on individuals for following their convictions (Figure 2.1). If formal institutions demand a high price from decision makers for acting in accordance with their convictions, such as losing professional position or prestige, then individuals are not likely to act honestly. This, of course, assumes that rapid cultural changes do not offer decision makers any other possibility to escape the formal constraints. In contrast, if the formal institutional price for revealing convictions is low, the convictions are likely to materialize in decisions as well, again holding all other factors constant (North 1992; Scott 1987).

The problem of obscuring information in response to institutional pressure

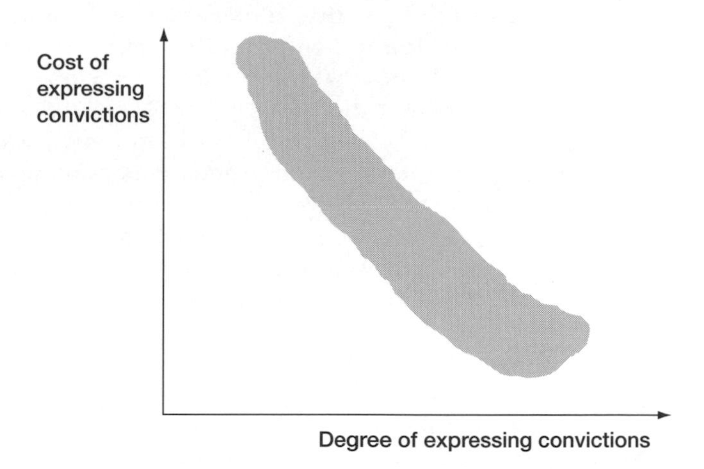

Figure 2.1 Institutional price of convictions.

is a serious one, because it questions the very idea underlying institutions. The *raison d'être* of institutions is that they reduce the uncertainties relating to social and economic transactions by increasing the transparency of information exchange (Berger and Luckmann 1967; Rutherford 1996). Pressures to change institutions are blatant, when incompatibilities have become so serious between what formal rules persuade decision makers to decide and what informal but popular professional convictions persuade them to believe in, that formal institutions have turned into perversions of the purpose they were originally intended to fulfil. This, according to the case evidence of this book, is indeed the problem in many contemporary environmental management controversies.

When tensions between informal and formal institutions do lead to changes in the formal institutions, the change usually occurs first in the lowest level of the hierarchy of institutions discussed earlier. Rules that guide the day-to-day operations of an environmental management system tend to change (or be changed) first. The last to change are rules that constitute the system (Ostrom 1994). The institutional reform recommendations of this book follow this prioritization. As will become evident in Chapter 8, the first priority is given to proposals to change the operational rules that determine the division of authority between regulatory interests and implementing interests, and collective choice rules that define the administrative lines of duty in the management of environmental technology. Answers will be sought to questions such as: how can an irrigation system be managed more reliably with fewer toxicity problems? Who should be responsible for the regulation of hazardous waste management? If you are in the business of waste disposal, should you also be in the business of waste reduction? The last in the order of priority, and also the most difficult to make operational in the form of clear design principles, are the recommendations to alter the constitutional rules that have to do with the relationship between long-term sustainability and short-term profit. The practical challenges here include: do we know what sustainability is? How might we find out? Is anybody in charge of long-term environmental policy and management? If not, who should be?

The preceding outline of the process of institutional change might be misconstrued as a claim that institutional change invariably leads to a dynamically stable system. Tensions between formal and informal rules might be seen as always finding a way of relieving themselves through changes in formal institutions. This is not the case. Economically inefficient and socially unsatisfactory systems have been shown to persist over long periods of time (North 1992; Rutherford 1996). The reasons can be traced to a mutually reinforcing feedback mechanism between the mental models of individual decision makers and the institutions within which they make decisions. Explication of the feedback is precisely what this work aims to do.

FEEDBACK BETWEEN EXPERT THINKING AND ENVIRONMENTAL INSTITUTIONS

Institutions persist because of a lock-in, or reinforcing feedback: the institutional framework determines what organizations come into existence and how they evolve, but the perceptions of individuals working in the organizations also influence how the institutional framework evolves (North 1992). A similar notion of feedback can be found in Berger and Luckmann's (1967) sociology of knowledge. They describe the dialectical process by which the members of a social group produce an institutional world that they begin to experience as an external objectivity, something other than a human product. This institutional world then acts back upon individuals through typified patterns of conduct shared by the members of the social group.

A particular version of this feedback was observed in the case studies of *Institutions in Environmental Management*. The policy makers and experts interviewed in the case studies frequently expressed the need to take into account the long-term future in environmental management, but the institutional framework within which they operated was a disincentive for them to develop and implement environmentally sustainable policies. Increasing returns on initial resource commitment in environmental technology and the complexity and uncertainty of long-term environmental issues reinforce the short-term oriented institutions. While the specific policy outcomes are different in each case study, the cognitive structure that the policy makers and experts share is cognitive dissonance, i.e. an inconsistency between individual decision makers' preferences (i.e. policies they think *should* guide decisions) and operating assumptions (i.e. policies they think *will* guide decisions) (Festinger 1962).

In the case studies the cognitive dissonance is between preferences emphasizing long-term sustainability and operating assumptions aiming at short-term profit. In California, for example, irrigation officials thought a drainwater conduit to the Pacific Ocean should be constructed to stop salt accumulation in the Central Valley in the long run. At the same time, they firmly believed the waste water outlet was not politically and economically feasible and would certainly not be built in the foreseeable future. In Finland, environmental regulators prioritized waste prevention as the top goal in hazardous waste management and recognized that it involved significant structural changes in industry in the long run. Yet their regulatory decisions were geared toward directing an increasing amount of hazardous waste to the nation's hazardous waste monopoly to guarantee its short-term profitability.

Cognitive dissonance is not merely the occurrence of conflict between the values of an individual official and the values on which the environmental policy of the agency employing the official is based. It is rather a conflict between two very different knowledges that officials have about themselves and their surroundings: one relating to the state of affairs an official would like to achieve in the society, the other to what the official considers achievable under existing constraints. The profit-oriented operating assumptions that dominate the US, European and Chinese societies pressure the officials there to reduce the dissonance by deciding on the basis of short-term economic considerations. This is also the way the theory on cognitive dissonance would predict them to behave (Festinger 1962; Hosking and Morley 1991). Put another way, the price – be it losing prestige, power or even position – that individual officials would have to pay for deciding according to their honest professional beliefs, is too high (North 1992).

Acting in accordance with the predominant assumptions of the surrounding society does not, however, erase the cognitive dissonance the officials experience. Quite the opposite. It exacerbates the conflict between what the officials believe in and what they know to be the necessary course of action. The cognitive dissonance individual decision makers experience and the profit-oriented institutions surrounding them are thus in a mutually reinforcing feedback. Institutional pressure to achieve short-term economic benefits exacerbates an individual's cognitive dissonance, which the individual then attempts to reduce by adhering to the institutional rules (Figure 2.2).

The feedback can also be expressed in terms of formal and informal institutions. Every interviewee in the four case studies of *Institutions in Environmental Management* was trapped by the same cognitive dilemma. This indicates that there exists a shared informal rule, which the interviewees have crafted in response to formal institutional demands. To summarize the

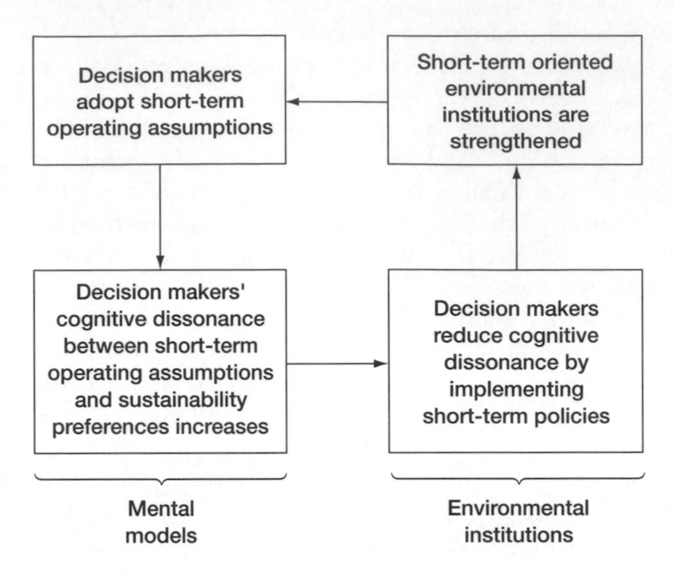

Figure 2.2 Feedback between formal environmental institutions and the mental models of experts.

results of the four case studies, that informal rule states that, whenever faced with an environmental management decision which at the most fundamental – constitutional – level involves a choice between short-term economic profit and long-term environmental sustainability, the decision maker should choose the former as the least costly alternative in professional terms, but continue to believe in the latter to avoid losing professional convictions altogether. This informal rule is a reaction to a set of formal institutional rules at constitutional, collective choice and operational levels. The constitutional rule in the sample countries is that national policies, environmental ones included, should in the final analysis be guided by the principle of maximizing the short-term economic well-being of the nation. The consequent collective choice rule stipulates that the different stages of an environmental management technology, beginning with the extraction of natural resources and ending with the disposal of waste in the environment, should be administered by one organization as a continuum of engineering operations, and without regard for the fact that the management of different stages of environmental technology have radically different time-scales. Finally, the formal operational rule dictates that long-term environmental policy should be administratively linked with short-term regulation and implementation of environmental management. The cognitively dissonant informal rule is a reaction to these formal institutions, but it also ensures closure of the feedback relationship between informal and formal rules:

decision makers adhering – if grudgingly – to the short-sighted imperatives also legitimize the existing institutional order.

Maintaining the uneasy balance between preferences and operating assumptions brings obvious benefits to environmental managers. It ensures stability in the way they define and tackle problems and in doing so protects them against co-optation, or even 'capture', by external groups (Scott 1987; Wilson 1989). Furthermore, since the issues of sustainability are so complex and uncertain, what the managers really want to do, but cannot, is not necessarily the right thing to do. Hence the call of this book for environmental institutions that would relieve the managers of their cognitive dilemma (by offering them a clear mandate in sustainability) while enabling them to maintain both organizational stability (through a diverse and adaptive organization) and the ability to cope with complexities and uncertainties (through organizational learning). These aspects will be elaborated further in Chapters 8, 9 and 10.

The unfortunate outcomes of the cognitive dilemma are slightly different in each case study. When the environmental issue is acute, as were the toxic compounds in Californian drainwater, policy makers focus on research and development to convince the general public that they are dealing with the problem. At the same time, they avoid long-term action that might inspire their critics to question the rationale of an economic activity that is the source of serious environmental problems. When the issue is not urgent, as in the case of Finland's waste management, policy makers turn away from the sustainability issue. Instead, they refine technologies and administrative procedures that ensure compliance with regulations, but also reinforce the wasteful industrial infrastructure.

ENVIRONMENTAL MANAGEMENT AND THE NEW AND OLD INSTITUTIONALISM

Much of the literature on institutional economics today is characterized by demarcation between the old and the new institutionalism. It is useful to position the case studies in this book and the argumentation about them in this debate. Central themes of the old institutionalism include the impact of new technology on institutional schemes and of property rights on economic transactions. New institutionalism focuses on, among other things, organizations, public choice processes, transaction costs and game theory. A dominant dividing line between the two traditions has been between holistic and individualistic approaches. Old institutionalists sometimes accuse new institutionalists of individualism, which in their view is a theoretically and empirically groundless attempt to explain social phenomena as a result of individual motivations and actions. New institutionalists in turn have dissented from the old institutionalists' holistic efforts to explain social processes in terms of institutional conditioning of individual behaviour (Rutherford 1996).

Although I refer to game theory and new institutionalism throughout *Institutions in Environmental Management*, the book does not unambiguously fall in either camp. Looking at the case studies in retrospect, they seem to me to be practical efforts to make the two traditions 'speak to each other' (Rutherford 1996: 181). Elements from both schools of thought are accommodated. This is less a conscious choice than the result of an interaction between empirical data and theoretical explanation that arises out of the pragmatic attempt to understand an environmental management problem and find socially practicable solutions to it. The argument is individualistic in the sense that mental models are the primary empirical data. It is holistic in the sense that the analysis focuses on aggregated individual models and their institutional constraints.

I will use a framework developed by Rutherford (1996) to show that the old and the new institutionalist views speak to each other in the analysis of the feedback between cognition and institutions. Rutherford relies on five perspectives to compare the new and the old institutionalism: anti-formalism versus formalism, holism versus individualism, rationality versus rule-following, evolution versus design, and issues of efficiency and reform.

The formalism debate stems from the inverse relationship, widely acknowledged among even the most ardent modellers, between the precision of a model and its ability to explain reality accurately. The greater its mathematical elegance, the more likely it is that the model loses the capacity of natural language to describe the richness of reality. Conversely, the truer the verbal description of what is going on in a particular environmental management issue, the less generally applicable that description is to other environmental problems (Ostrom 1994). Although some cognitive mapping methods I have applied in this study, such as the Bayesian network analysis, are highly quantitative, the explanation of the feedback between mental models and environmental institutions is descriptive. There are good reasons for this. First, quantification of the factors of the descriptive model, such as cognitive dissonance, would be impossible due to the qualitative nature of thematic interview data. Second, the case studies are too few to attempt to formalize any generalizable quantitative models. Finally, there is the ruthlessly pragmatic justification of a policy analyst: why bother to quantify if you can get reasonable policy recommendations without it?

The individualism versus holism debate is about the relevant level of analysis needed to explain social processes. Is the relevant level of analysis the individual or some aggregation of individuals? To explain what is going on in modern environmental management issues in terms of a feedback between individual mental models and environmental institutions is to draw the arrow of causation explicitly both ways between the two levels of analysis. On the one hand, the short-term and profit oriented environmental management institutions are the result of individual decision makers trying to reduce their cognitive dissonance by adhering to the predominant rules of the society. On the other hand, the cognitively dissonant behaviour of individual decision makers is only strengthened by the short-term oriented institutions. This book thus contains an explanation of how individuals think and act, given institutional entities, *and* of how institutional entities function, given individual patterns of thought and action. Both lines of thinking were evident in the interviews. In the Californian case, for example, one expert felt that since individual 'farmers are fighting each other and are not willing to make sacrifices', farming in the state had turned into 'a dog eat dog world' (interview 5). But another expert saw it the other way around: 'Ideally, one could think of some kind of subsidy or incentive programme to solve the problem', which, however, was unlikely, since 'we have the kind of government that emphasizes individual choice', and therefore 'farmers who have marginal agricultural lands cannot bear more costs and will be driven out of business' (interview 1).

The question in the rationality versus rule-following issue is whether individuals are maximizers of their individual utility or followers of social rules and regulations. The decision makers in the case studies clearly choose a middle ground between the two theoretical positions. They are what March and Simon (1994) have termed 'satisficing' individuals under conditions of bounded rationality. Since they cannot, under existing institutional constraints, select what they consider to be the optimal alternatives that would meet the goals of long-term environmental management, they opt for alternatives that, while causing a significant degree of cognitive dissonance, are none the less still minimally satisfactory. The cases are illustrations of social situations in which decision makers who act rationally under existing institutional boundary conditions at the same time increase the likelihood of political, social and bureaucratic tensions over issues of sustainable development and long-term environmental management. A Colorado water agency official, for example, expressed 'a vested interest in maintaining the *status quo*', because, 'if something were to come up that says there will be no more irrigated agriculture [in the Arkansas River valley], [w]e would have no customers for our inventory, and companies go broke if they cannot sell their inventory'. Yet the same official was 'not too sure that irrigated agriculture as such is going to continue as a dominant feature here. [O]ur agriculture here is getting pretty old, because the water rights here have been for irrigation since the 1850s. [. . .] There really has never been, except for the Egyptians, I guess, a society that has endured that depended on irrigated agriculture' (interview 4).

On the issue of institutional evolution versus design, this book, once again, takes the middle ground. My argument is clearly oriented toward design, because the outcome of each case study is a set of institutional design principles. Yet I recognize that the conditions that enable serious consideration of institutional design are the result of an evolutionary process, and provide, in essence, an infrequent window of opportunity for design. It just so happens that at this particular time the biophysical and institutional environments in the case study regions have co-evolved into a state where environmental decision makers and experts are convinced that there exist significant incompatibilities between the industrial production system and the ecosystem, and hold ideas about how to resolve or otherwise manage the incompatibilities. They just lack the autonomy to put the ideas into action – most vividly expressed by Chinese environmental experts, who believed economic development would always win over environmental protection, because the supreme political authority of the time, Deng Xiaoping, had allegedly stated in public, 'Economy first, environment second' (interviews 1 and 5). At this stage in the evolution of environmental institutions, a potential for influencing the path of future evolution exists, and that potential, I think, could be realized with better, more opportune design.

But what is the desirable institutional design? How do we define institutions that are efficient, let alone guarantee the social good? And how can we be sure our design intervention, and not a blind evolutionary process, will be the best guarantor of efficiency or social good? The position underlying the institutional design recommendations of this study is that the efficiency and the social good of a particular institutional design are defined by the stakeholders who have to tolerate or enjoy the institutions. The criteria for determining the desirability of an institutional set-up are therefore socially defined, and subject to constant argumentation and redefinition. The only thing planners can wish to do is stimulate the creation of institutions that enable the stakeholders to search and find an acceptable consensus on environmental issues, instead of continuing to redefine and disagree on them. This position also means that the design recommendations cannot be too specific, because the beneficiaries of the design will be future generations, whose preferences and definitions of the social good are, for all intents and purposes, unknown. Instead, the design recommendations should be considered as proposals for trial in a complex and dynamic evolutionary process. The belief that the learning process of trial and error might be worthwhile stems from careful analysis of documentary material and interviews in the case studies, which indicates that the stakeholders themselves perceive that they are in the midst of serious environmental challenges in need of remedial action.

The description of the feedback between mental models and environmental institutions is based on case studies. The primary empirical data in the case studies were interviews with experts and decision makers working in government agencies, private corporations, environmental interest groups, consultancies, universities and research centres. Before turning to the case studies, I will discuss in Chapter 3 the methodology of the thematic interviews and the cognitive mapping exercise with which I extracted the mental models from the interview data and codified them as issue networks and causal networks.

3

FINDING THE INSTITUTIONAL RULES

To conceive of institutions as rules of the social game links institutions closely with the way individuals think. Rules reflect cause-and-effect relationships. But causality is also a fundamental organizing principle of individual thinking. Individuals tend to recount their professional opinions with narratives that have a causal structure (Bower and Morrow 1990; Tversky and Kahneman 1982). When asked to describe the environmental challenges they were facing in the broadest sense, the interviewees in the case studies typically provided a rich account of the most important problems, potential solutions to the problems and, most importantly, the institutional constraints to the solutions. The causal mental model is thus an individual's interpretation of the institutional rules that constrain his or her decisions. This makes the study of institutions also a study of cognition.

The main institutional argument of *Institutions in Environmental Management* states that the actions of environmental decision makers are largely determined by a feedback between institutions and the mental models of these decision makers. The predominant institutional order in the case study societies prioritizes short-term economics over long-term sustainability, which persuades environmental decision makers to adopt cognitively dissonant mental models. While thinking that long-term environmental concerns should guide their decisions, the decision makers believe that short-term economic concerns will in the end determine which policies will be implemented. The more the formal institutions dominate the actions of decision makers, the stronger the cognitive dissonance they experience; but the stronger their cognitive dissonance, the more decision makers try to reduce it by adhering to the existing institutional order.

This chapter describes the methodologies used in the case studies to analyse the feedback. Using the stories that decision makers and experts tell about an environmental issue as the database for institutional analysis provides valuable information for institutional design. First, the interviewees are capable of identifying the significant institutional dimensions of the issue, precisely because they are the experts. Here, for example, is a Finnish expert with a premonition of the state of affairs that I would in my later analysis call environmental corporatism, or the institutionalized mixing of conflicting environmental interests:

> So that's a kind of a special feature, which I think is a special feature of all small countries, that at a certain level everybody always knows everybody. It is very advantageous in daily management, but whenever you're dealing with this conflict between short- and long-run goals, well, then it has disadvantages. Because the issues never enter more comprehensive deliberations.
>
> (interview 4)

Second, and more importantly, analysing as an aggregated whole the mental models of many experts grappling with the same environmental issue permits the outside analyst to see something the experts themselves may well not see on their own. The individually conceived mental models of the environmental controversy form the building blocks for determining the interviewees' socially constructed reality (or realities) of the issue. Circular arguments found in the case studies, both at the individual and group levels, constitute the most intriguing aspect of this social reality. A Californian water quality regulator, for example, was obviously not sure where to put the burden of responsibility in the state's agricultural drainage management. At one point in the interview, the regulator stated that the regional agency regulating water quality in the state's Central Valley, the so-called Regional Board, 'does not want to [regulate drainage] because it sees the problem not as a regulatory problem but more as a resource management problem, which would be best taken care of at the local level'. Yet, at another point during the interview, the same expert lamented that, although the local level 'has a lot of good people to take up the leadership', they are 'not willing to do it' and 'understand neither the physical problem nor regulation and cannot respond to regulatory actions taken by the Regional Board' (interview 5).

Finally, since the starting points of the empirical analyses in the book are individual mental models, the institutional recommendations emerging from the analyses are closely linked with important contextual and individual factors of decision making. The mental models that can be observed in individual narratives are complex causal networks containing both normative and factual statements about the particular environmental management decision the individual is dealing with at the time of the interview. This is how a Colorado water official expressed his value laden yet factually grounded expert judgement of sensitive water issues in the American West during the confidential interview:

> Well, that's where you can make a case about . . . [looks out of the door] nobody here in agriculture, nobody has a – you have a

> tape-recorder on but I'm not afraid of yours . . . marginally productive lands. Once again, what I spoke of this couple from Rutgers [University] who advocate the buffalo commons [converting western US farmland to buffalo pasture], maybe that's really going to happen. If you look at the big agricultural picture that the USDA [US Department of Agriculture] has painted for this country for a number of years, they've tried very hard to keep the small farmer in business, they've wanted food to be cheap, they've propped up the commodities. But what that's meant is that the marginally productive areas have stayed in business. [. . .] Here we are paying people not to grow in the Midwest, and we're paying people to grow out here. There's a whole question there that I wonder about. People are raising alfalfa even in the Colorado River Basin. People grow it so easily in the Midwest, and we pay them not to raise it. The system's at a point of breaking down somewhere, I think.
>
> (interview 9)

The next section provides a theoretical background for the cognitive mapping of expert interviews. I will then describe the sampling of interviewees and the interview procedures, followed by an elaboration of the network analytical methods with which I investigated the interviews. The chapter concludes with a discussion of the validity and reliability of problem networks as a way of codifying the mental models of experts.

COGNITIVE MAPPING

All the case studies began in an atmosphere of uncertainty and complexity. In the California agricultural drainage case (see Chapter 5), each decade since the 1950s had witnessed a new governmental report on the drainage problem and its possible solutions, none of which had been implemented due to economic, political and environmental uncertainties and disputes. What could be said or done constructively about a policy problem, in which the only empirically verifiable certainty appeared to be that within the next ten years another task force would find the issues to be technically, economically and environmentally poorly understood, complex and therefore in need of further research? In the Finnish waste management case (see Chapter 6), environmental policy makers were under pressure to present visionary strategies on waste prevention, while barely coping with the reality that the landfill capacity of several municipalities was rapidly coming to an end with few or no waste disposal alternatives in sight.

Since past research on the issues had not realized viable solutions, the starting position in the case analyses was to avoid the preconceptions formed about the problem by previous analysts. Instead, the point of departure was to solicit considerable input and differing views from the practitioners, planners, regulators and chief critics of environmental management – in short, the key decision makers and experts on the subject, both in the government and in society at large. After all, the single most important constraint on any analysis of a politically charged issue is the need to ensure that the plans take into account the often differing perspectives of the major actors in the controversy (Nelkin 1984). Put another way, many environmental policy issues have become so uncertain, complex and polarized that just about the only things left for the analyst to study are the different stories with which policy makers and their opponents articulate the uncertainty, complexity and polarization (Roe 1994).

The stories of individual decision makers and experts are data for cognitive mapping. Underlying the individual stories are the mental models with which decision makers in environmental management authorize and rationalize their decisions. The mental models centre around interrelated issues or problems that can be coded as issue networks or causal problem networks, respectively. The focus of analysis in each case study was the structure, substance and institutional context of the mental models of key environmental policy makers. Experts and decision makers in an issue give advice and make decisions on the basis of mental models they have formulated about past and future development. It is not the empirically verified reality that determines their decisions, as many of them would like to claim, but rather what they perceive to be the reality (Berger and Luckmann 1967; Weick 1969). Furthermore, policy makers must frequently act on incomplete information and process the information they do receive through mental models that can lead to choices that reinforce existing institutions as the only source of stability and certainty – however inadequate or far from optimal, in any sense of the term, those institutions might be (North 1992). Such was the *modus operandi* of the central California water district manager, who claimed to be unaware of any toxicity problems in the district at a time when the state's wildlife researchers were reporting evidence of toxic concentrations of selenium in the food chains of three of that district's evaporation ponds, and significant adverse biological effects in one of those ponds (San Joaquin Valley Drainage Program 1989: 2–26–2–31, Tab.2–7). Open admission of toxicity problems at the height of other controversies over equally polarizing issues, such as agricultural crop and water subsidies and migrant farm labour conditions, would have posed much too great a threat for the stability of the state's water institutions.

For these reasons, I do not treat individual scenarios as elements of a plausible theory of the sequence of events the future might hold. Instead, I consider individual scenarios as equal elements of a collective, and often internally inconsistent, model of and for the future. Analysis of the factual, attitudinal and logical inconsistencies of the collective model indicates how institutions constrain long-term policy decisions and what institutional changes might relax the constraints (more on cognitive mapping, see Eden 1992; Eden *et al.* 1979).

It is important to note that my analytical approach is very different from such collective problem solving techniques as brainstorming or the Delphi technique. Brainstorming comprises several techniques aimed at discovering new ideas on the basis of intuitive thinking and achieving consensus by a number of people (Clark 1958; Jantsch 1967; Osborn 1963). The Delphi technique in effect combines several brainstorming rounds by interrogating individuals sequentially, with an attempt to avoid interference of the psychological factors that reduce the value of conventional brainstorming sessions (Bell 1967; Gordon and Helmer 1964; Jantsch 1967). In these techniques, the expert testimony is subject to criticism and comment from other participants. This book's research allowed no such feedback. The difference is not coincidental. Analytically most interesting here are not the unexpected ideas that might emerge in a brainstorming session or the staunch opinions defended in a round-table debate, but rather the original and uncontested yet well-considered and honest expert opinions expressed during a confidential interview.

Despite the differences between them, the case studies are all firmly grounded in the theoretical framework of collective problem solving. The unifying premise is that the providers of expert opinion form an integrated social subgroup and articulate issues with a vocabulary understood by each subgroup member (Jantsch 1967; McCarthy 1985; Saussure 1966). The members may, of course, differ in their interpretations of the problem, but can in principle always locate the sources of such differences through discourse (Nelkin 1984). Indeed, as critical social theorists remind us, discourse itself presupposes the existence of consensus:

> The very act of participating in a discourse, of attempting discursively to come to an agreement about the truth of a problematic statement or the correctness of a problematic norm, carries with it the supposition that a genuine agreement is possible. If we did not suppose that a justified consensus were possible and could in some way be distinguished from a false consensus, then the very meaning of discourse, indeed of speech, would be called into question.
>
> (McCarthy 1975: xvi)

The social subgroup of environmental experts and decision makers therefore holds a socially conceived perception of reality with sometimes similar, at other times dissimilar views of the environmental problem and its potential solutions (Berger and Luckmann 1967). Analysis of such similarities and dissimilarities in expert beliefs and opinions forms the foundation for discovering a sociopolitical redefinition of the environmental management problem.

That its focus is on beliefs and opinions rather than empirically verifiable biophysical facts makes the analysis no less empirical. Key analytical questions are how the social reality is constructed out of individual beliefs and how that reality and individual beliefs constitute each other (Berger and Luckmann 1967). The data in the case studies are the expert beliefs and opinions, and their analysis produced the institutional insights and propositions described in Chapter 2. The analysis is therefore fundamentally inductive, i.e. the hypothesis is inferred from the empirical evidence supporting it. The situation in which no hypothesis or full-blown causal theory, but rather a theoretical framework, guides the analysis, is not new to organization and policy sciences. As Selznick admits in his ground-breaking study on the Tennessee Valley Authority, 'while approaching his materials within a guiding frame of reference [of organization theory], the author was not committed by this framework to any special hypothesis about the actual events', and emphasizes that the abandonment of his initial working hypothesis after fieldwork was a 'major illuminating notion' (Selznick 1984: 251 n.6).

EXPERT INTERVIEWS

The empirical data in each case study come from interviews with key decision makers and experts on the environmental management issue in question. I extended the social constructionist logic, which allows the interviewees themselves to define the key environmental issues, to the selection of interviewees. I used snowball sampling, in which those already interviewed identify who else they think should be interviewed. First, I selected a set of core interviewees on the basis of a survey of the literature on the environmental management issue and discussions with officials in key organizations dealing with the issue (typically the agency initiating the study and its research arm). I chose subsequent interviewees through snowballing by asking each core interviewee who he or she thought were the most significant experts and decision makers on the environmental issue (Burt 1982). In the Californian and Finnish case studies, however, snowballing was not taken to the theoretical end where additional interviews are conducted until the last interviewee can only suggest experts already mentioned by previous interviewees (Goodman 1961). Instead, an interviewed expert had to have been mentioned by at least a given number of previously interviewed experts (two in the Californian case study, three in the Finnish one). This limited the sample size in each to a maximum of about twenty interviewees, which is feasible considering the workload involved in the gathering, transcription and codification of qualitative interview material.

Limiting the sample size raises the question of sample representation. In the Californian case study, an assessment of the adequacy of sample size

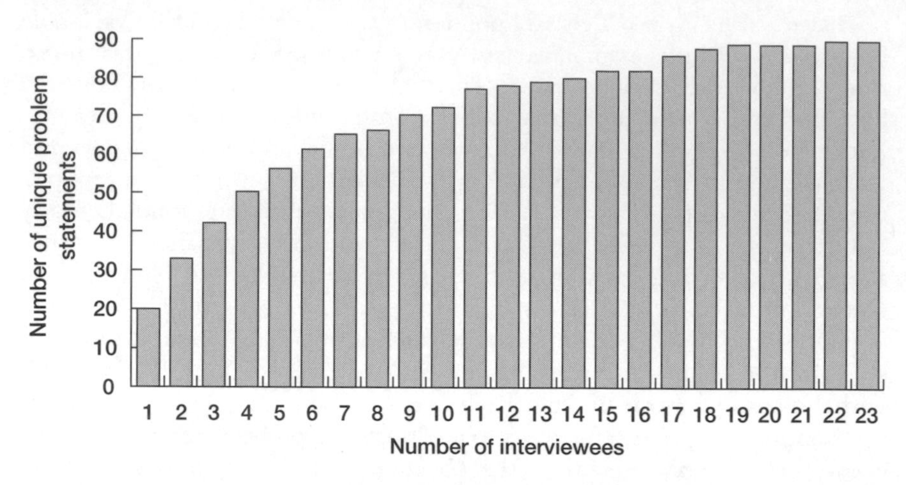

Figure 3.1 Cumulative distribution of problems in the Californian case study.

indicated that the twenty-three interviews identified what the parties involved in the controversy perceived to be the major environmental problems. The cumulative frequency distribution curve of problems mentioned in the interviews (Figure 3.1) shows that the number of new problems added by individual interviews (when the interviews were ordered so as to maximize the incremental increase in the number of previously unmentioned problems) begins to taper off after about the ninth interview and that fifteen interviews already captured 90 per cent of the problems (Hukkinen 1990). Additional confidence in sample size adequacy comes from the fact that in all cases the last few interviewees, when informed of the organizational affiliation of experts interviewed thus far, commented on the 'good coverage' of the interviewee sample (for reasons of confidentiality, the identities of experts interviewed could, of course, not be revealed). Finally, contextual evidence I collected in each case study from related literature, i.e. sources other than expert interviews, corroborates the conclusions of the interview analysis, which increases my confidence that the interview sample was adequate.

I classified the interviewees into interest groups on the basis of previous analyses of the environmental issue and the classification the interviewees themselves utilized during the interview. The groupings have a clear correspondence with the elements of organizations frequently referred to in organization theory. The *technical core* can be identified, and it is made of groups such as the agricultural community and planners in the Californian case study, or the public waste management agencies and private waste management firms in the Finnish one. In all case studies, environmental regulators are the most important element of the *task environment* of the agencies and firms dealing with the environmental management issue. Finally, environmental activist groups represent the most influential interests in the environmental management agencies' *institutional environment* (Perrow 1970; Scott 1987; Thompson 1967).

The interview sessions, which usually lasted from one to two hours, were, as a rule, conducted with one interviewee at a time. The interviews were thematic, qualitative and informal: I asked the interviewees a few loosely structured and open-ended questions about the environmental management problem at hand and potential solutions to it. In the Colorado and Finnish case studies, I tape-recorded the interviews; in the Californian and Chinese ones, I took notes in writing. The data recording technique was a compromise between the conflicting pressures to obtain undistorted information about the environmental management issue and to record that information with precision and accuracy. In the Californian case study the drainage issue was deemed to have reached a level of political volatility that might have made the interviewees reluctant to speak on tape. In China, this reluctance appeared to be culturally ingrained. I transcribed the handwritten notes and the tape-recordings into narrative form, usually within a few days of the interview.

Interviews in the Chinese case study had to be conducted with some exceptions to what has been described above. The only acceptable format to the Chinese interviewees turned out to be group interview. This was an adaptation of methodology to the cultural setting, probably at some cost to the comparability of data. After all, the idea in individual interviews is to obtain mental models of experts without interference from their peers. But it was a compromise worth making in China, because the alternative would have been no data at all.

What is more, there are good empirical reasons to argue that the group interviews were no compromise after all. To view individuals as the main actors in a society is deeply rooted in the tradition only of some disciplines, such as neoclassical economics. Institutional economists and cultural anthropologists remind us, however, that over the long haul, collective or affiliative forms of organization have dominated human cultures – and often still do – whereas individualism and individual liberty are associated mainly with modernity, democracy and prosperity (Bennett 1996; Ostrom 1994). And Chinese culture is certainly known for its ancient traditions and deeply embedded patterns of collective organization, both in official centres of power and in informal family and clan connections (Christiansen and Rai 1996). A social subgroup of experts instead of an individual expert may therefore have been the only relevant level for obtaining the mental models in the Chinese context.

Another problem in China was language. Having no knowledge of Mandarin, I had to rely completely on interpreters during the interviews, which inevitably resulted in some loss of information I might have considered important for my research but the interpreters did not. On one occasion the environmental manager of a state enterprise went into a lively fifteen-minute Mandarin monologue in response to my request to describe the main environmental problems at the factory. My interpreter's English version of the monologue was, 'We have no major environmental problems at the factory.'

Yet the similarities between the Chinese case study and the Western ones were great enough to justify the presentation of the case studies in one volume. The method of interviewee sampling was the same across the cases. Furthermore, a clear causal structure could also be observed in the arguments presented during the Chinese group sessions, allowing the same network analytical methods to be used as in the Western case studies. Finally, to facilitate comparability, I paid particular attention to the Chinese cultural context by comparing English and Mandarin terminology in environmental policy and management, and investigating the institutions, organizations and managerial practices of environmental protection in China.

Some inadequacies and biases inevitably result from the methods of sampling and interviewing. The research approach is openly elitist: the interviewees are either decision makers with significant leverage on the outcome of environmental policy and management, or experts in key advisory positions to the decision makers. The assumption, therefore, is that these individuals are well-tuned enough to the ongoing environmental debate to consider the weak signals from interests not represented in the sample. However, there are no assurances of this. Weak signals with significant influence on institutional evolution or design may have been omitted. Furthermore, even though the loosely structured interviewee narratives describe well the causal processes operating in an environmental controversy, more formalized interview methods would give a clearer quantifiable picture of the environmental management issue (Babbie 1986; Bailey 1982; Fielding and Fielding 1986). The timing of the interviews may also have biased the experts' descriptions of the environmental problem. Had the interviews been conducted some years later, for example, when the results of additional studies would have become available, individual models of the environmental problem would probably have been much more specific. The hypotheses inferred from the interviews none the less appear just as valid, especially in light of the supporting contextual evidence I present in each case study from a time both before and after the interviews.

NETWORK ANALYSIS

The analytical approach of viewing individual mental models as equally valid parts of a socially constructed environmental issue arose in all cases from preliminary investigation of the interview data. Agricultural drainage management in California is a case in point. Since the early 1950s, a considerable number of policy analyses and technical and economic feasibility studies have been conducted to find ways of managing and disposing of agricultural drainwater in California's Central Valley. By the end of 1989, the latest governmental task force, the San Joaquin Valley Drainage Program, had spent an estimated 50 million US dollars on research of drainage-related problems and solutions, but had not found technically or economically feasible and politically acceptable solutions.

The inability of the San Joaquin Valley Drainage Program to reach a consensus even at the fundamental level of problem definition can be gauged from the interviews I conducted with the twenty-three experts representing the key interest groups in the debate. The interviews were originally to provide the basis for identifying the list of salient drainage alternatives and assessing their technical and economic feasibility. This task, like its predecessors of the past four decades, turned out to be impossible. The exercise assumed that some kind of consensus existed over problem definition. In reality, the interviewees gave a number of different and often conflicting descriptions of the drainage problem. With only twenty-three interviews, no less than ninety different drainage-related problem statements were identified. Very few problems were mentioned by more than two interviewees: fifty-two of the ninety problem statements were recorded only once or twice, and only eight problems were mentioned by seven or more experts. No set of problems could therefore be singled out as the primary target for remedial action. More important, when the interviewees did state the same problem, their perceptions of causality often differed widely: what was a cause from one expert's viewpoint proved to be an effect from another's. Only nine problems were classified as just one type of problem (e.g. as either a cause or an effect) by all who mentioned them, and none of them was included among the eight most frequently mentioned (Hukkinen *et al.* 1990).

I found indications of similar scattering of opinion in the other case studies as well, although I did not conduct preliminary analyses at the level of detail they were done in California. In Colorado, nine interviews produced 109 distinctly different descriptions of the water management issue; in Finland, twenty-four interviewees identified 282 unique problem statements about long-term waste management; and in China I observed fifteen problem statements in the six interviews (note that in China this includes only problem statements concerning environmental policy and regulation; I made no attempt to code the rich descriptions of environmental engineering challenges as problem statements).

In the Californian and other case studies in this book, the inability to apply conventional policy analytical tools to compare environmental management options triggered a search for an alternative analytical approach.

The stories and scenarios of the interviewees were analysed with the objective to identify the controversy's underlying set of beliefs and premises about relevant environmental problems and their relationships. Each interview is considered not as a test of an externally constructed model of relationships claimed to be operating in a controversy that is taken as given, but as an equally valid element of a larger cognitive map from which the environmental issue is constructed.

Two types of relationship characterize the interviewee narratives. First, the interviewees frequently cast environmental issues in the form of oppositions between a general description of the environmental problem and a specific rebuttal of that problem (or, alternatively, a general prescription on how to solve the environmental problem and a specific restriction on how not to proceed toward the general goal). This is how interviewee 5 expressed the opposition in the Colorado irrigation management case:

> The system has evolved and become accustomed to a pattern of return flows [i.e. irrigation water not used by plants]. My predecessor [. . .] has a saying, and it is true: 'One man's waste is another man's water supply in the Arkansas.' So the system has adjusted to accommodate whatever practices are in place. But I'm not sure those practices are the most efficient from a water distribution standpoint. And from a water quality standpoint, I think that there must be a problem associated with the practices of water use that generate return flows. I understand that the water quality deteriorates by a factor of two just in a twenty-mile stretch between Fowler and La Junta. The return flows that are occurring in that area are degrading water quality, there must be a correlation there.

In other words, the interviewee perceives the problem to be that irrigation practices along the Arkansas River of Colorado are not the most efficient, either from a water distribution or from a water quality standpoint. On the other hand, the interviewee does not believe this is much of a problem, since the state's water management system has adjusted to accommodate whatever practices are in place.

Second, definitions of the environmental problem were expressed in causally related arguments specifying a problematic sequence of phenomena. The following excerpt from an interview with a Finnish waste management expert illustrates the flow of causal argumentation:

INTERVIEWEE 1: I think one threat is that Finland's scarce resources will be tied to heavy systems and that other options will in this way be closed out of consideration. Capital will be tied to solutions with a long time-span. I think Ekokem [Finland's centralized hazardous waste plant] is already one example of this. Ekokem was designed with excess treatment capacity. Well, this then led to a situation where even waste that didn't necessarily have to go to Ekokem had to be collected there. One such fight was over waste oil.

JH: Was this the one where they [the regulators] didn't allow waste incineration elsewhere, because . . . ?

INTERVIEWEE 1: Exactly. And then they ended up trying to find – artificially, in fact, like looking for a needle from a haystack – what harmful effects such incineration would have if it occurred some place other than Ekokem. When they had rammed this one through, they then found themselves in a situation where Ekokem could no longer incinerate all the waste that arrived there. After which they began to forward the waste to be incinerated at precisely those plants where they had at first banned it. This really takes away the credibility of all this policy very badly.

The asymmetry between causalities and oppositions is the major dividing line between the two respective networks that are used to analyse the interviews, namely, causal networks and issue networks. In causal networks, interviewee arguments are nodes and causal relationships between the arguments are arrows. In issue networks, relationships between issues are more contextual and lack causal direction: issues are nodes and contextual relationships between issues are links. The Californian, Finnish and Chinese case studies were analysed with causal networks, the Colorado one with issue networks.

That causalities and oppositions should become the foci of analysing loosely structured narratives comes as no surprise. Linguists have observed that causality and contrast are the fundamental connective relationships whenever a speaker presents his or her personal opinion in natural language: they are used in everyday discourse as well as in highly specified argumentations and scientific investigations (Dijk 1977; Rudolph 1988). Research in cognitive psychology also supports the notion of constructing networks of causal connections among events in a narrative (Bower and Morrow 1990; Tversky and Kahneman 1982).

In the Colorado case study (see Chapter 4), which illustrates an easy-to-use heuristic method applied to a relatively small sample of interviews, the analytical focus is solely on issue-based relationships. In the rest of the case studies the relationships teased from the interview narratives are both causal and issue linkages between perceived environmental problems. Since analysis of issue-based relationships is really a more general case of the analysis of causal relationships, I will focus on the latter in the following.

The interviewees typically described environmental management problems in complex, causal narratives. A directed node-and-link graph, or problem network, conveniently represents the causal dependencies between problems expressed during the interview. Figure 3.2 shows how the earlier

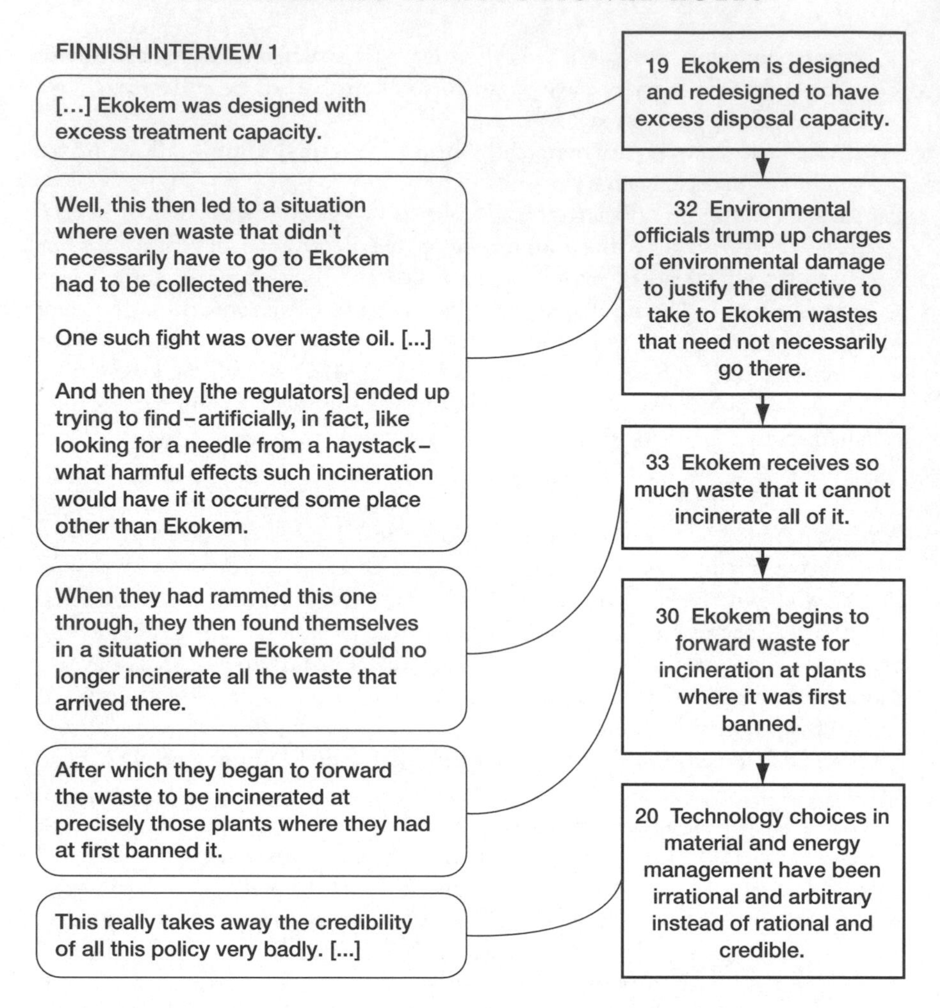

Figure 3.2 Codification of a Finnish waste management expert's narrative transcript as problem network.

quotation from the narrative transcript of a Finnish waste management expert was coded as a problem network. Nodes in the graph are problem statements, which are sentences describing a single aspect of an environmental management problem. Links in the graph represent those parts of an interviewee narrative that indicate a causal relationship between two problem statements (Harary *et al.* 1965; Hukkinen 1993a; Hukkinen *et al.* 1990; Pearl 1988; Wilson 1974). Since each interviewee has his or her own description of the environmental controversy, a unique problem network can be constructed from the narrative of each interview. Graph theory provides a consistent way of coding as a problem network the complex web of interrelated problems identified in individual narratives.

Network representation is useful in two ways. First, it makes explicit the cause-and-effect relationships that otherwise might go unnoticed in the rich narratives. Second, it allows the analyst to consider the implications of several individual models in the aggregate. Network aggregation in its simplest form proceeds as follows: if interviewee A argues that problem statement 1 leads to problem statement 2 (1 → 2), while interviewee B feels that problem statement 2 leads to problem statement 3 (2 → 3), then the aggregated problem network of interviewees A and B becomes 1 → 2 → 3. In other words, problem statement 2, which is a terminal problem for interviewee A and an initial problem for interviewee B, becomes a transfer problem after aggregation.

Network aggregation has several advantages. First, it reveals to what extent the lack of convergence over problems and causality is due to contradictory or circular arguments at individual or intra-group level, rather than conflicting arguments between interest groups. Second, aggregation at the inter-group level identifies causal relationships that only become clear when the views of the controversy's participants are considered together. Third, it makes potential sources of future conflict between interest groups clearer, since the aggregation exercise is the closest approximation of the debate that might arise should all the major participants sit around a table and argue the points they raised individually in the interviews. Such conflict is obviously only a potential one, since a simple aggregation exercise is insensitive to the interaction effect of individuals modifying their privately held positions when speaking publicly. Lastly, aggregation at individual and group levels enables representation of the environmental conflict as a combination of individual and group perceptions that imposes system-wide problems across the same individuals and groups. As the detailed case descriptions will illustrate, the aggregated problem networks are the most visible and concrete signs of the feedback between individual mental models and their systemic institutional constraints.

Network aggregation reveals causal relationships that an interviewee alone does not necessarily appreciate. A particularly intriguing network configuration is the directed cycle, or loop. In the simplest loop, problem 1 is perceived to be leading to problem 2, which in turn leads back to problem 1 (1 ↔ 2). The Californian case study offers an example of an actual loop in the debate over agricultural drainage management: the US Bureau of Reclamation's abrogation of its contracts with local districts to provide return flow services has led these districts to threaten suits against the Bureau, which in turn has made the Bureau all the more rigid in its stance of not honouring the contracts.

Loops are central to long-term policy planning. They are circular arguments, which blur the distinction between cause and effect that is a prerequisite in the design of any policy. Since the networks in the case studies are composed of expert beliefs about the unfolding of past or future environmental problems, the loops are destabilizing positive feedbacks rather than equilibrating negative ones. Interviewees think environmental management problems are getting increasingly worse instead of remaining the same (on positive and negative feedback, see Bertalanffy 1968; Gleick 1988; Jantsch 1985).

Another focus of network analysis consists of the terminal paths resulting from the loops (i.e. chains of problems that originate in loops and end in terminal problems). To give an example, some experts believe that the above-mentioned loop concerning the US Bureau of Reclamation's abrogation of drainage contracts has enabled many politicians to take advantage of the situation by raising public concerns over drainage, which in turn has worsened the NIMBY ('not-in-my-backyard') syndrome against all public works facilities and led to the terminal impasse, in which none of the potential drainage management solutions can be implemented. Loops together with terminal paths emanating from them are the most persistent problem configurations, because loops sustain themselves, and terminal paths persist by virtue of the loops. There are, of course, initial problem paths leading to loops, but doing away with them would affect neither the loops nor the consequent terminal paths. The Bayesian probability analysis of the problem networks in the Californian case study (see Chapter 5) presents a more quantitative justification for focusing on loops and terminal paths.

VALIDITY AND RELIABILITY OF PROBLEM NETWORKS

How well do the causal problem networks describe the social reality of the experts? And how closely do the problem networks correspond with actual biophysical phenomena underlying the environmental issues of the case studies? The first question aims to assess the validity and reliability of problem networks as representations of what the interviewees said. It is a key question, because it probes the empirical soundness of the social constructionist approach of this book. But we should also be concerned with the 'external' validity and reliability of problem networks, because the interviewees frequently refer to biophysical facts in their problem descriptions. Comparing what the experts say the biophysical facts are with what can be learned of those facts from other sources can provide important insights as to why the experts argue as they do. I will discuss both types of validity and reliability.

The transformation from the problem descriptions of an individual's narrative transcript into the problem statements of a problem network is not clear-cut. The problem statement is a category of sentences describing an aspect of the environmental problem, expressed in at least one narrative transcript. However, it is a category of only roughly similar sentences. As would be expected, different interviewees often use different expressions and sentence structures to identify the same problem. The problem link, on the other hand, has an even wider variety of possible counterparts in the narrative transcript. Sometimes a causal connection is expressed explicitly with connectives such as 'because', 'since', 'therefore', 'consequently' and 'as a result'. At other times, however, the causal connections are expressed in subtle contextual relations in the text, just as the literature on text generation and text grammars would predict (Dijk 1977; McKeown 1985; Rudolph 1988).

In a sense, variation in narrative problem descriptions classified under the same problem statement and variation in causal expressions coded as the same problem link are analogous to variation in any measurement of social or physical phenomena. The problem statement can be thought of as a discrete variable with as many different states as there are unique problem statements mentioned in the interviews. The problem link is a binary variable with two possible states: the link either exists or does not exist. Criteria would therefore have to be specified to estimate both the validity and the reliability of using the problem statement and problem link as the chief variables representing the narrative transcripts' problem descriptions. Validity requires checking, first, that the variable actually measures the concept of interest, and not some other concept; and second, that the concept is being measured accurately by the variable. Reliability, on the other hand, refers to the consistency of the measurement (Bailey 1982).

In problem networks, the purpose of validity determinations is to make sure that the problem statement and the problem link actually represent the environmental problem descriptions (and not some other type of descriptions) of the narrative transcripts, and that they do it accurately. Reliability, on the other hand, requires that a given problem statement or problem link has consistently similar counterparts of individual problem descriptions or their causal relationships, respectively, in the narrative transcripts. From the above discussion on the nature of problem statements and problem links it is clear that making such validity and reliability verifications presents some difficulties in problem networks.

The first difficulty, which relates to the validity of problem networks, has to do with the analyst's interpretation of what the interviewees said about environmental management. Even if we assume – and I realize it is assuming a lot – that the interviewees were honest, there remains the possibility that the analyst misunderstood what was being said. Were this to be the case, the problem statements and problem links could not be held representative of the problem descriptions of the interviewees. In the Californian drainage management case and the Finnish waste management case, I made an attempt to prevent this by submitting a draft report of the findings of problem analysis to staff members of the key agencies. Comments by staff members were subsequently taken into account in refining the results of the analysis.

The second difficulty encountered in the validation of problem networks has to do with the accuracy of problem statements and links in representing the problem descriptions of the interviews. To increase the accuracy of problem statements, I used key words and expressions similar to those found in the narrative transcripts when formulating the statements. Accuracy of problem links is a less complex issue, due to their binary nature: all that is needed for a description of causality to be coded as a problem link is the existence of a connective word or context. However, as I will point out soon, the ambiguity in the relationship between problem network elements and their narrative transcript counterparts in turn raises issues of reliability.

Third, validity also includes consideration of how representative the set of problem statements is in depicting the 'problem world' of the interviewees. I analysed this only in the Californian case study with a cumulative distribution of the number of problems mentioned in the twenty-three interviews. This analysis indicates that the twenty-three interviews have covered sufficiently the major problems perceived by the key experts in the San Joaquin Valley drainage debate (see Figure 3.1).

Fourth, and much as a result of the difficulties relating to accuracy, the reliability of problem networks as representations of the interviewees' problem world cannot be determined rigorously. The reliability criterion just stated requires that problem statements and links have consistently similar counterparts in the narrative transcripts. In a strict sense this is impossible, since some problem statements and links were expressed in completely unique terms or contexts, thereby invalidating any claims to consistent similarities. Such shortcomings in verifying the reliability of problem networks as representations of narrative transcripts stem from the often-noted richness and irregularity of texts and the consequent difficulty in creating grammars for them (Beaugrande 1980; Greimas 1987; Riffaterre 1983). Acknowledging those difficulties, this research relied on the conviction that the language used by the interviewees, despite its specific dissimilarities, was none the less an expression of a Saussurian 'social contract' among environmental decision makers (Saussure 1966: 77–8). Indeed, the environmental decision makers and experts interviewed in the case studies do form a highly socialized and compact community through their frequent formal and informal contacts with each other in planning, day-to-day operation, regulatory procedures, public hearings and public participation arrangements. This, after all, is why they referred to each other during the snowball sampling procedure. I therefore consider expert interviews a reliable basis for drawing inferences about the environmental problem that are collectively understandable and meaningful – though not necessarily acceptable – for the decision making community as a whole.

Figure 3.3 illustrates the correspondence between interview transcripts

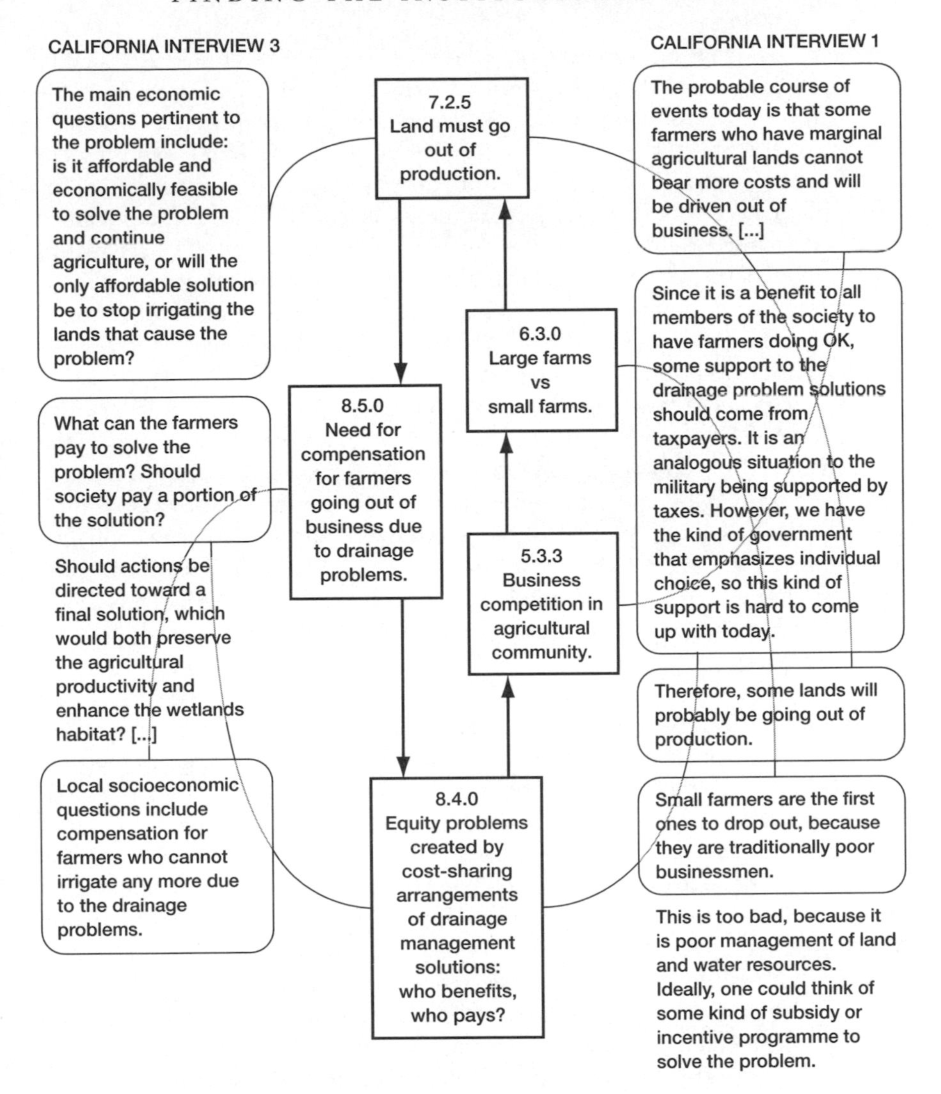

Figure 3.3 Correspondence between the irrigation bureaucracy's interview transcripts and terminal loop in the Californian case study.

and one of the loops of California's irrigation bureaucracy. Figure 3.2 does the same with interview data from the Finnish waste management case study. Appendix 1 provides more detailed examples of the transformations that took place in the coding process of narrative transcripts into issue networks in the Colorado case study and into problem statements and links in the Californian case study. The purpose of Figures 3.2 and 3.3 and Appendix 1 is not to prove the validity or reliability of creating issue and problem networks, only to give the reader an idea of how closely attached to or far removed from the coded networks the narrative transcripts are.

In addition to assessing how validly and reliably the problem networks represent the socially constructed reality of environmental experts, the analyst should also be concerned about the 'external' validity and reliability of problem networks. By this I mean how closely the causal relationships specified in the problem networks correspond with the biophysical reality that the experts are dealing with, such as irrigation-induced toxicity, hazardous waste quantities or environmental engineering applications. Although the focus of *Institutions in Environmental Management*, for reasons outlined earlier, is what the experts perceive to be the reality, external validity is none the less a relevant question. After all, the biophysical reality is the main reference point for the issues that the experts discussed in the interviews. A marked discrepancy between the social and biophysical realities, as expressed by the Californian water manager reluctant to acknowledge obvious toxicity problems in evaporation ponds, indicates political manoeuvring by an expert grappling with significant institutional constraints, and is therefore of major interest for the analyses in this book (lack of factual knowledge is a less likely reason for the discrepancy, because the interviewees were experts).

To validate the assertion that problem statements and links represent actual biophysical phenomena would require independent verification of the factual basis of the stated problems. But in many cases the data needed for this verification can only be obtained from the same experts whose statements were to be verified in the first place. Furthermore, a problem statement may refer to biophysical facts yet in substance be an expert's value judgement about a problematic state of affairs, making independent verification of its truth value logically impossible (Habermas 1975; Hume 1978; Popper 1977; Simon 1964). None the less, whenever logically and practically possible, I ascertained the empirical merits of the problem networks with information obtained from related literature. Putting the socially constructed realities in the context of the biophysical reality in this way illuminated the institutional complexities in all four cases.

It is now time to turn to the empirical case studies on which the methods of cognitive mapping were applied and the institutional proposition is based. The next chapter presents the first case study, which deals with institutional issues of water management in the dry areas of southern Colorado and illustrates the utilization of a heuristic cognitive mapping method on a relatively small interview sample.

Part II

CASE STUDIES ON MODERN ENVIRONMENTAL MANAGEMENT

4

INSTITUTIONAL DISTORTION OF WATER QUALITY MODELLING IN SOUTHERN COLORADO

The first case study is a story of failed efforts to understand and solve one of the oldest environmental problems known to humankind with one of the most promising modern environmental management technologies. Increased interest in nonpoint source (i.e. from geographically diffuse sources) pollution control in the US has revived discussion about the management and regulation of agricultural drainage over the past couple of decades. Adequate drainage is necessary to maintain irrigated agriculture over time. The purpose of drainage management, also known as return flow management, is to ensure adequate leaching of water and salts from the soil. Without drainage agriculture is impossible as soil becomes waterlogged and saline (National Research Council 1989; San Joaquin Valley Drainage Program 1990; US Committee on Irrigation and Drainage 1987). Computerized water quality modelling has played a central part in efforts to solve large-scale agricultural water and salt management problems in the Arkansas River Basin of southern Colorado since the 1970s. The use of models in the design of policy and management intensified in the early 1990s. Computer models can provide a credible scientific basis for the management and regulation of irrigation-induced drainage by formalizing for management purposes the causal relationships believed to be underlying the observed natural and social phenomena.

At the same time, any effort to control return flow on the basis of a model must gain acceptance in the eyes of relevant interest groups and in the context of prevailing institutions. The Colorado case study is a warning to those who believe that the only thing needed for successful environmental policy implementation is an adequate amount of objective and reliable science to support decision making. In today's volatile water management issues, science is embedded in institutions, making modelling not just a scientific, but a political act as well.

In the Arkansas River Basin, the organizational and political context of irrigated agriculture distorted the application of a computerized drainage model. Cognitive mapping produced a qualitative expert system, i.e. a set of expert rules on how the drainage management system in the Arkansas River Basin would react to policy intervention. I used the expert system to evaluate modelling in the design of irrigation return flow control policy for the river basin. The main finding is that concern for the economic viability of the basin's irrigated agriculture and the effects of regulatory action there have prevented effective use of modelling in drainage policy.

At a more general level, inadequate model application in the Colorado case study illustrates the institutional proposition of *Institutions in Environmental Management*. While the Arkansas River Basin's drainage experts thought long-term environmental concerns should guide decisions about return flow policy, the dominant institutions persuaded them to prioritize and implement policies that secured the short-term economic viability of irrigated agriculture. The more the formal institutions dominated an individual expert's decisions on return flow management, the stronger that individual's dissonance between preferences and formal institutions. But the stronger the cognitive dissonance, the more the expert tried to reduce it by making every effort to 'stabilize' the return flow management system with adherence to formal irrigation institutions. In the Arkansas River Basin these dynamics could be found on two levels of social rule: the constitutional rule and the collective choice rule (see Chapter 2). At the constitutional level, the case study illustrates the impacts of short-term profit orientation in agribusiness in the US, in both federal and state operations. At the collective choice level, the case study shows that administrative merging of irrigation and drainage responsibility is a disincentive to far-sighted agricultural management.

Methodologically the Colorado case study shows that cognitive mapping need not involve extensive interviewee samples with sophisticated methods of analysis. What I report here is an easy-to-use heuristic cognitive mapping exercise on just nine interviews, which is none the less capable of bringing out in great detail the intricacies of both the institutions of agricultural management and the mental models of irrigation experts.

WATER MANAGEMENT IN COLORADO'S ARKANSAS RIVER BASIN

The Arkansas River connects Colorado's climatic and geographic extremes (Figure 4.1). The highest peaks in the 25,400 square mile (65,800 square kilometres) river basin reach above 14,000 feet (4,300 metres) and receive an average precipitation of over 40 inches (1,020 millimetres). In contrast, average precipitation falls below 10 inches (250 millimetres) in the eastern plains of the river basin, which is at an elevation of 3,350 feet (1,020 metres) at the Colorado–Kansas state line. Most of the precipitation in the

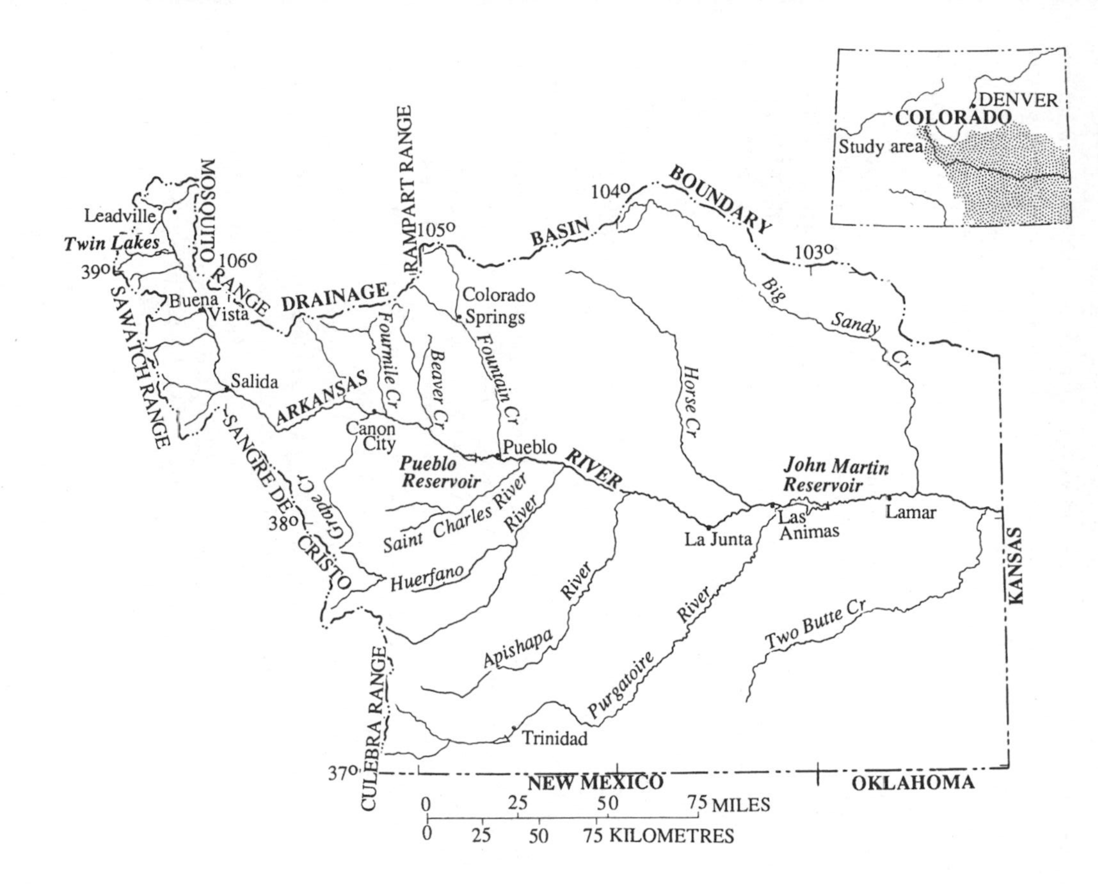

Figure 4.1 The Arkansas River Basin in Colorado (Burns 1989: 3)

arid eastern plains comes during intense summer storms. In the mountains, precipitation occurs during the winter, but is stored in a thick snowpack which melts during the spring and summer (Abbott 1985).

Extreme natural variations in streamflow have always determined the pattern of water use along the Arkansas River. In the 1840s, individual farmers could meet their need for water with direct diversions from the river. By the 1880s, mutual irrigation companies had fully appropriated the river and its tributaries during normal or average years. The first off-channel reservoirs to level off streamflow variations were built in the 1890s, and were followed in the next century by larger reservoirs constructed by the US Army Corps of Engineers (USCE) and the US Bureau of Reclamation (USBR) (Figure 4.1). Groundwater pumping from the river's alluvial aquifer began in the 1940s, increased dramatically during the 1950s, and slowed down after the 1960s when it started to have negative impacts on the water supply of the river itself. Approximately 20 per cent of the water applied to crops was pumped from groundwater during 1941–65 (Abbott 1985; Burns 1989; Cain 1985).

The Arkansas River is the water supply of about 305,000 acres (123,000 hectares) of irrigated farmland below Pueblo reservoir. Cantaloupe and sugar beet were originally responsible for the basin's agricultural development. Current crops include melons, corn, sorghum, alfalfa, beans, wheat, onions, tomatoes, cucumbers, aubergine, cabbage, chiles, grapes, cherries, raspberries, apples, peas and squash. Surface irrigation is the only irrigation method. On-farm water-use efficiency (i.e. the ratio of the amount of water infiltrated and stored in the soil's root zone to the amount of water applied to the field) averages 40–50 per cent. The river basin is heavily over-appropriated, and water shortages can reach 40 per cent in an average water year (Cain 1985; Sutherland and Knapp 1988).

The river salinity increases considerably going downstream. Part of the salinity originates in naturally occurring salt deposits in the river basin, part of it is due to the concentrating effects of water use and reuse by irrigators. The maximum salinity values in the river range from 770 to 5,100 parts per million of total dissolved solids (ppm TDS), with the highest levels in the lowest reaches of the stream (Cain 1985; Sutherland and Knapp 1988). There is some inconclusive evidence that salinity in the alluvial aquifer may have reached a long-term dynamic equilibrium below Pueblo (Cain 1987). According to expert interviews conducted in this study, up to 70 per cent of the irrigated land has subsurface tile drains to alleviate soil salinization and waterlogging. The current effectiveness of the drainage systems is questionable, however, because many of them were built as early as the 1930s.

Efforts to control saline irrigation return flow discharges to the Arkansas River are hindered by the fact that return flows from the upstream irrigators are an important water supply for the downstream irrigators. The notion of return flow as a water supply is the very essence of Colorado's water institutions and relationships between water agencies. As a result, water and return flow management in the state is conditioned by a complex web of physical, social, legal and economic interdependencies.

Colorado water law in particular strengthens the interdependencies of water management. Eastern states in the US follow the so-called riparian doctrine of water rights, which states that anyone who owns property along a stream has a right to the water in that stream. In contrast, western states have adopted different versions of the so-called prior appropriation doctrine, which gives the person who gets to the stream first the superior right to the water over those who get there later. 'First in time, first in right' is the often-quoted summary of prior appropriation. Where California recognizes both riparian and appropriative water rights, Colorado interprets prior appropriation strictly by repudiating the riparian water right as unsuitable for semi-arid conditions. The interpretation facilitates intricate linkages among different water users, because appropriative rights may be traded and switched up- and downstream. Prior appropriation can also encourage waste, since claimants who establish a right to water but do not use it can forfeit their right – also known as the 'use it or lose it' principle (Dunning 1982; Powledge 1983).

Irrigated agriculture in Colorado's Arkansas River Basin is currently at a critical conjuncture. The once-prosperous irrigation-based agricultural economy has over the past five decades suffered a number of diverse setbacks. Gradual salinization of irrigation water supply has forced downstream irrigators to switch to salt-tolerant but less profitable crops. Downturns in agricultural economy and national unwillingness to continue restrictions on cane sugar led to the disappearance of the basin's famous sugar beet industry in the 1960s. Economic problems now threaten irrigated production as a whole. An increasing number of valley farmers are taking advantage of Colorado's purely appropriative water rights law and selling their water to the state's fast-growing cities. An estimated 136,000 irrigated acres (55,000 hectares), 72 per cent of which are prime cropland, have the potential for being removed from agricultural production (Sutherland and Knapp 1988). At the time of the case study, speculation about the possible inclusion of irrigation return flows in the federal National Pollutant Discharge Elimination System (NPDES) water quality permit system also contributed to the widespread perception that the valley's agriculture is threatened. Another issue of concern in the early 1990s was the outcome of the lawsuit by Kansas against Colorado over the latter's alleged violation of the Arkansas River Compact (Commissioners for Colorado 1948), which could potentially have reduced the irrigators' water supply in Colorado's Arkansas River Basin. The Supreme Court has previously made decisions on water litigation between Kansas and Colorado in 1907 and 1943. In both cases the Court used the economic reasoning that the benefit to Colorado far outweighed the detriment to Kansas. However, the Court did point out that a time could come when Colorado would appropriate so much water that Kansas could insist on its equitable share (Radosevich *et al.* 1976).

Indeed, most experts interviewed for this study had resigned themselves to believing that irrigation will eventually disappear from the basin. At the same time, the experts were careful not to let this belief affect their short-term goal of securing the year-to-year viability of farming in the basin. The result was a deep-rooted unwillingness to change the delicately balanced *status quo* of irrigation management. The experts could claim some merit for this operating assumption. Agriculture still seemed to thrive in many areas of the basin. The ecological and social disruptions caused by water transfers to cities did not speak well for the feasibility of the proposed alternatives to irrigated agriculture. But as the following sections show, the operating assumption also undermined efforts to use modelling in the design of a long-term irrigation return flow control policy for the Arkansas River Basin.

EXPERT INTERVIEWS ON DRAINAGE MODELLING

Several reasons led me to choose Colorado's Arkansas River Basin as a case study of the effects of institutions on return flow modelling. The Arkansas River Basin has been the subject of intensive modelling since the early 1970s (Burns 1988, 1989; Konikow and Bredehoeft 1974; Konikow and Person 1985; Lefkoff and Gorelick 1990a, 1990b). An extensive data base of river flows and salinities covers more than four decades. According to a 1991 review I conducted of drainage models in the United States, California's San Joaquin River Basin (see Chapter 5) and Colorado's Arkansas River Basin were the only sites in the US where a drainage model was being used to design return flow policy (Hukkinen 1991b). Finally, at the time of the case study the Soil Conservation Service (SCS) was in the process of modifying the US Geological Survey's (USGS) Interactive–Accounting Model (IAM) for use in the design of return flow management for the Patterson Hollow project area, which consists of 90,000 acres (36,000 hectares) located on the south side of the Arkansas River between La Junta and Pueblo (Figure 4.1) (Soil Conservation Service 1990).

The IAM (or the Burns model; see Table 4.2, the interviewees frequently refer to this model after its designer) simulates streamflow, water quality, and water supply operations in the entire Arkansas River Basin in Colorado. In the model the Arkansas River and its tributaries are conceptualized as a network of nodes. The model's regression equations compute flow from drainage areas between the model nodes by using a time series of independent variables, such as snowpack, precipitation or gauged streamflow. Salinity (TDS) concentrations and loads are computed from regression equations at each model node by using streamflow as the independent variable. Streamflow and salinity loads are then routed downstream (Burns 1988, 1989).

This study set out to investigate the institutions within which the IAM would be used in the Arkansas River Basin. I conducted interviews in April 1991 with nine irrigation experts in agencies that research, plan, manage or regulate Colorado's water operations. Two experts were with the SCS, two with the USGS, one with the USBR, one with the Colorado State Engineer's Office, one with the Southeast Colorado Water Conservancy District, one with the Region 8 of the US Environmental Protection Agency (EPA), and one with the Colorado State Department of Health's Water Quality Control Division.

The interviews were informal and loosely structured. The experts were asked a number of open-ended questions about irrigation return flow control problems in the Arkansas River Basin and the use of models to solve those problems. Whenever the interviewee had knowledge about the topic, the interview also focused on the application of the IAM. The loosely structured interview format allowed the experts to include those contextual issues in their narratives that they felt were the most critical in the application of models for return flow control.

COGNITIVE MAPPING

The experts' accounts of return flow and model application problems are comprehensive and rich in detail. A unifying theme runs through all interviews. The experts are reluctant to characterize irrigation return flow management in the Arkansas River Basin as having any serious 'problems', saying it is at most a 'challenge' (interview 3). They do not even want to talk about instituting a return flow policy to meet the challenge, because they feel the system is 'all kind of hooked together' (interview 4) and has reached an ill-defined 'optimal' stage (interview 7) by 'adjusting to accommodate whatever practices are in place' (interview 5). Return flow quality problems have been 'compensated for by the agricultural entities over the hundred years that the system has been irrigated', and people have 'learned to live with the salinity' (interview 6).

Every return flow issue the experts discuss reflects the general notion of irrigation return flow management as a challenging but stabilized system to be treated with caution. An expert 'stabilizes' the return flow problem he or she is discussing with a cognitively dissonant rebuttal: when considered in the context of other return flow issues, the problem (1) is not really a problem; (2) should not be solved; (3) cannot be solved; or (4) would cause other problems if solved. Conversely, when discussing a potential solution to a return flow problem, an expert rebuts the solution by stating that, in the context of other drainage issues, the solution (1) is not really a solution; (2) should not be implemented; (3) cannot be implemented; or (4) would result in other problems if it were implemented. An example of a rebutted problem is the following pair of statements: 'Over-irrigation is part of the irrigation return flow problem. But over-irrigation also ensures a water supply for the downstream water users.' A rebuttal of a solution is 'More dams should be built on the Arkansas River, because there is not enough storage for the irrigated land. But the feasibility of building more dams is questionable in today's political climate.' Since the pairs of statements are neither clear-cut problems nor solutions but rather dispute both, they will be termed issues. Cognitive dissonance is common among the experts, since a total of 109 different issues could be identified in just nine interviews. The issues can be categorized into eight groups based on their substance (Table 4.1).

Issues in the first category focus on the linkages between irrigated agriculture and accumulating soil and water salinity. Issues in the second category describe the effects of using return flow as a water supply in the river basin. The third group of issues outlines the mechanisms by which Colorado's water laws often encourage the waste of water. Water use and reuse in the river basin raise issues between upstream and downstream water

Table 4.1 Classification of issues in the Arkansas River Basin's irrigation return flow debate

Issue category	*Description*
1 IRRIGATION AND SALINITY	Linkages between irrigated agriculture and accumulating soil and water salinity in the Arkansas River Basin
2 RETURN FLOW AS WATER SUPPLY	Effects of using return flow as water supply in the Arkansas River Basin
3 COLORADO WATER LAW AND WATER WASTE	Mechanisms by which Colorado's water laws encourage wasteful use of water
4 UPSTREAM VERSUS DOWNSTREAM	Conflicts between upstream and downstream water users resulting from water reuse and salinity increase in the Arkansas River Basin
5 WATER TRANSFERS FROM RURAL TO URBAN AREAS	Conflicts resulting from water sales from rural to urban areas in Colorado
6 MODEL USE	Applications of irrigation return flow models in the Arkansas River Basin
7 REGULATION	Issues arising from regulatory control of irrigation return flows in the Arkansas River Basin and elsewhere in the US
8 RESPONSIBLE PARTIES IN RETURN FLOW CONTROL	Attribution of responsiblity for return flow management in the Arkansas River Basin

users, which fall under the fourth category. Colorado's purely appropriative water law permits water transfers, which has resulted in controversial water sales from rural to urban areas, as outlined in issues belonging to category 5. Issues in the sixth category describe the various applications of irrigation return flow models. The seventh category contains issues relating to regulatory control of return flows. Finally, issues in the eighth category deal with the question of who should be considered responsible for return flow management in the basin.

The experts frequently stated causal or contextual linkages between the return flow issues they mentioned. For example, the above-mentioned proposition to build more dams on the Arkansas River (issue 4) has linkages with the following issues (as explained in Chapter 3, I have formulated the issues using key words and expressions similar to those found in the narrative transcripts of the interviews):

> 50. Better irrigation scheduling to provide water on demand would ease the return flow problem in the Arkansas River. But the area is water short and there is not enough storage capacity along the river to provide the flexibility required for irrigation scheduling.
>
> 92. Construction of Pueblo reservoir eased the water shortage in the Arkansas River by increasing storage capacity. But at the same time farmers lost the sealing benefits of silty water, because silt settles out in the reservoir.
>
> 93. When farmers lost the sealing benefits of silty water after dams were built on the Arkansas River, they tried water-conserving irrigation techniques, like sprinklers. But there the salinity becomes a problem, because it clogs the pipelines.
>
> 123. Because of the purely appropriative water rights in Colorado, farmers use their allotment of water irrespective of precipitation, and are reluctant to store water for future use. But the water management system in the Arkansas River has grown to operate so well under the current rules, that any technical change is going to affect the functioning of the overall system.
>
> 154. Some people may propose solving the irrigation drainage problem by building more dams to provide more water for dilution. But that would only encourage more ineffective use of water and lead to more of an agricultural welfare economy.

The causal and contextual linkages connect an issue both with questions specific to return flow control and with questions related to the institutional environment of return flow control. The above-mentioned issue 4, for example, which had to do with the construction of more dams, is linked to such drainage-specific questions as improving return flow control through irrigation scheduling (issue 50 above), and the trade-offs between providing water storage and managing return flows (issues 92 and 93). But issue 4 also has institutional ramifications, both directly (because it refers to 'today's political climate') and indirectly (issue 123 talks about the 'functioning of the overall system' of water management in the river basin, and issue 154 about the effects of water use patterns on 'agricultural welfare economy').

The objective of the interview analysis was to describe as accurately as possible the equilibrated return flow management system to which the experts frequently referred. The linked drainage issues that an expert describes make up a mental model with which the expert explains drainage phenomena. The equilibrated drainage management system can be constructed by aggregating the mental models of all interviewees. The resulting collective cognitive map is an 'expert system', because the return flow issues described by the interviewees are also expert rules on how the drainage management system in the Arkansas River Basin would react to policy intervention. Obviously, the expert system is purely qualitative, since the interviewees were not asked to estimate the strength of the relationships between the issues. None the less, the system can simulate decision making processes in the river basin's return flow management and crudely predict the fate of future return flow policy proposals.

The overall structure of the collective cognitive map is presented in Figure 4.2. The 109 return flow issues are grouped into the above-mentioned eight substantial categories. Through its component issues, each category has causal and contextual linkages with other categories and with the institutional environment. To find out the effects of a return flow control proposal on the river basin's overall water management system, the proposal needs first to be identified with one of the substantial categories. Once the proposal is categorized, return flow issues directly related to the proposal are identified. These issues are either institutional issues or drainage-specific issues, and refer again to other issues. These references can be followed up further as needed. When all linkages have been investigated, a qualitative expert estimate of the feasibility of a return flow policy proposal has been obtained.

The cognitive map was used to investigate the expert community's response to the SCS's Patterson Hollow project. The chief purpose of the project is to increase irrigation efficiency as a way of controlling irrigation return flows. Less water applied on the fields would pick up a lower amount of salts from the soil, which would decrease salt load to the Arkansas River and the alluvial groundwater aquifer. The project's scientific credibility rests on the IAM. Testing the objectives of the project against the decision making principles that leading experts on

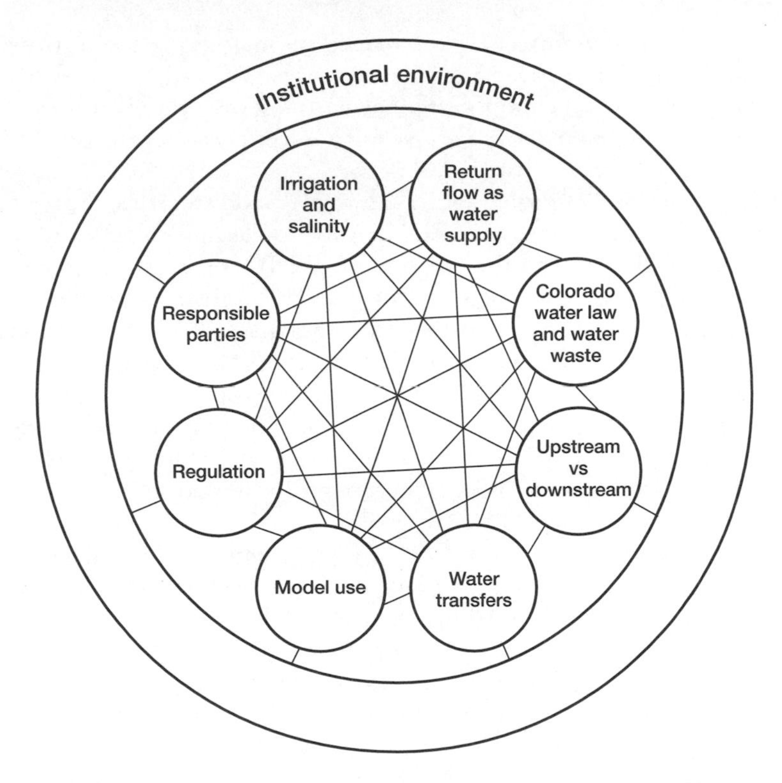

Figure 4.2 Structure of the collective cognitive map of Arkansas River Basin's irrigation experts.

the river basin's water management revealed during the interviews illustrates how organizational and political realities constrain the experts' use of the model.

STATUS QUO AND THE MISUSE OF A MODEL

The proposition to improve irrigation efficiency relates to issues in category 2 (return flow as water supply), which is therefore the starting point for analysing how the nine experts collectively respond to the proposition. The analysis consists of identifying issues relevant to the proposition, and using this set of issues as the qualitative data for detailing what effects the proposition would have on the overall system of drainage management in the Arkansas River Basin. First, the drainage management response is obtained by identifying the drainage-specific issues with which the proposition to increase irrigation efficiency has linkages (Table 4.2). Then, the response of the institutional environment is discovered by identifying the institutional issues with which the proposition has linkages. Table 4.2 focuses only on the primary impacts the proposition has on the seven other issue categories; secondary and higher order impacts resulting from changes triggered by the primary impacts are not listed.

Analysis of the Patterson Hollow project indicates that institutional factors inhibit the SCS from reaping any modelling benefits from the application of the IAM in the Arkansas River Basin. In general, computer models of hydrologic processes can be used to evaluate problems, design remedial strategies, conceptualize flow processes, guide the collection of data and improve the quality of decision making (National Research Council 1990). In the Arkansas River Basin, each of these potential benefits is lost. First, modelling confuses the drainage problem and the design of remedial strategies by focusing attention solely on technical factors instead of the river basin's central issues of economic viability and environmental regulation. Second, the IAM is incapable of conceptualizing the relevant flow processes without support from a more detailed model, which, however, the SCS has no plans for using. Finally, modelling cannot guide data collection or provide additional information for decision making because the SCS's reluctance to pinpoint the sources of salinity in the basin secures farmer support for on-farm management projects such as the Patterson Hollow. I will elaborate each of these points by relying on the institutional and drainage-specific issues in Table 4.2.

Application of the IAM in the Arkansas River Basin takes place in a context which return flow experts characterize as a gradually degenerating irrigation-based agricultural economy (institutional issues 66, 175 and 176 in Table 4.2). Threats against the survival of the basin's irrigated agriculture come from several directions: salinity build-up in the river and the agricultural lands surrounding it, which eventually makes cultivation impossible (institutional issues 66 and 153); water quality regulation by state or federal officials, which might make the already marginal agricultural economy unprofitable (drainage-specific issues 76 and 83 and institutional issues 60 and 135); and water transfers from the agricultural areas to Colorado's rapidly growing cities, which thus far have led to an erosion of both the physical landscape and the social infrastructure in rural areas (as indicated by many higher order linkages that the proposition to increase irrigation efficiency has with issues in category 5).

At the same time, the experts seem to have accepted the threatening state of affairs as the best of possible worlds. Some believe that salinity in the river system is not accumulating, but has in fact reached a long-term dynamic equilibrium. To them, salt levels are not increasing in the long run, and irrigators have learned to live with the high salinity that is present (institutional issues 70 and 124). Others think that salts are accumulating, but that any action to reduce upstream salt load to the river would cause

Project proposal: the Patterson Hollow project will increase irrigation efficiency through the control of irrigation return flows. Less water applied on the fields will pick up less salts from the soil, which will decrease salt load to the Arkansas River and the alluvial groundwater aquifer.

Expert community's response

Issue category	*Drainage–specific issues*	*Institutional issues*
1 IRRIGATION AND SALINITY	50 Better irrigation scheduling to provide water on demand would ease the return flow problem in the Arkansas River. But the area is water short and there is not enough storage capacity along the river to provide the flexibility required for irrigation scheduling.	66 Serious conflicts over irrigation return flow are unavoidable. But irrigated agriculture in the Arkansas River Valley is comparatively old anyway, and is not likely to continue as a dominant feature there because of salt build-up.
		115 The SCS on-farm project may have an effect on salt loads to the Arkansas River. But there is a chance that the river has not yet reached an equilibrium in terms of salinity, which would mean that the SCS is aiming at a moving target.
		153 The direction that the USDA is taking in their return flow programmes is to improve irrigation efficiency, cut the amount of water that irrigators are taking out of the river and bring to the crop only what's necessary to make it grow. But these measures may lead to a worsening of the salt accumulation problem in the root zone.
		175 Much of the USA's food supply depends on western irrigated agriculture, which is threatened by return flow problems. But the western farmers don't seem to worry about that, and sometimes even accept the downfall of irrigated agriculture.
		176 Return flow problems threaten the survival of irrigated agriculture in the western states. But maybe these marginally productive lands should be allowed to go out of irrigated production, and maybe the nation should concentrate on maintaining the good farmland of the Midwest.

Table 4.2 (Continued)

Issue category	*Drainage–specific issues*	*Institutional issues*
2 RETURN FLOW AS WATER SUPPLY	(Linkages with all issues in the category)	60 Solving the return flow problem might entail increasing the farmers' consumptive use of the river water. But since the river is water short, a lot of irrigated acres would have to go out of production as a result.
		70 Irrigation practices along the Arkansas are not the most efficient, either from a water distribution or from a water quality standpoint. But the system has adjusted to accommodate whatever practices are in place.
		111 Water quality control is another issue in improving irrigation return flow management. But water in the Arkansas River Basin just isn't of very good quality, and the only alternative to not having a quality problem is not using the water.
		124 One of the proposed solutions to the irrigation return flow problem is to institute water conservation practices on farms. But when the water management system depends on waste and leakage, conservation might have negative effects on the functioning of the overall system.
3 COLORADO WATER LAW AND WATER WASTE	22 According to Colorado water law, water saved through best management practices cannot be sold, which is not really an incentive for farmers to divert less water for irrigation. But since the water saved belongs to the next junior appropriator and therefore stays in the river system, it is not really lost.	81 Water users will at times take water just for the sake of exercising their priority, regardless of their need for water. But they have the legal right to do so, and they have a very strong perception of the sanctity of their right to divert water.
	48 Some of the irrigation return flow problems could be eased by convincing farmers to use less water. But irrigators have interpreted Colorado's prior appropriation doctrine to mean 'use it or lose it', which encourages over-irrigation.	110 Water quantity control is one issue in improving irrigation return flow management. But the system has reacted to the way it's worked for a hundred years – it has reached an optimum, and any attempt to change water quantity would just open up a huge water rights issue that is not easily solvable.

Table 4.2 (Continued)

Issue category	*Drainage–specific issues*	*Institutional issues*
	65 Colorado water laws are not clear on the reuse of return flows for irrigation. But return flow reuse is already a widespread practice in the Arkansas River Basin, which means that any return flow policy potentially opens up a number of unknown legal questions.	123 Because of the purely appropriative water rights in Colorado, farmers use their allotment of water irrespective of precipitation, and are reluctant to store water for future use. But the water management system in the Arkansas River has grown to operate so well under the current rules, that any technical change is going to affect the functioning of the overall system.
	82 If there is within a ditch company[a] a water need in one area and no need in another, the company will probably make water available to both. But such waste of water is necessary to keep things on an even keel internally, because the companies are answerable to their shareholders in distributing water in an equitable fashion.	137 Water use efficiency should be improved to manage irrigation return flows better. But since Colorado water law has created the need for return flow, landowners are reluctant to change their agricultural practices.
4 UPSTREAM VERSUS DOWNSTREAM	38 There's a challenge associated with the management of irrigation return flows in the Arkansas River Basin. But eliminating return flows completely would deprive downstream water users of their water supply, because the Arkansas River is heavily over-appropriated.	42 Downstream irrigators in the Arkansas River Basin have a considerably worse water quality than the upstream irrigators. But the lower river water users do not blame the upper river users for the poor water quality, because they realize that without the return flows resulting from the over-irrigation of the upper river they would not get any water.
	56 Over-irrigation is part of the irrigation return flow problem. But over-irrigation also ensures a water supply for the downstream water users.	90 Downstream farmers know that upstream farming has an effect on river salinity. But they've learned to live with the salinity and don't blame the upstream irrigators.

Table 4.2 (Continued)

Issue category	*Drainage–specific issues*	*Institutional issues*
	59 Solving the irrigation return flow problem might require eliminating all return flows to the Arkansas River. But this would mean that irrigators would have to increase their consumptive use of the river water, because so many irrigators depend on return flows for their water supply.	
5 WATER TRANSFERS FROM RURAL TO URBAN AREAS	(No linkages)	(No linkages)
6 MODEL USE	32 The Arkansas River Compact lawsuit between Kansas and Colorado had a chilling effect on the use of the Burns model,[b] because people are afraid of even asking water quantity-related questions until the lawsuit is resolved. But the lawsuit will probably not affect SCS's use of the model, because water quality is not an issue in the Compact.	14 Landowners are not interested in using the Burns model, because they are only interested in on-farm measures that can increase their crop productivity or decrease their operational cost. But the model could be used by governmental agencies to illustrate potential off-site benefits of the Patterson Hollow Project and to increase the likelihood of funding.
	52 Models may cause conflict among farmers if they show that on-farm management practices would increase the upstream irrigators' water use efficiency but decrease the amount of return flow to run in the downstream ditch. But models as such should not be blamed for the conflicts, because they only give the information needed in decision making.	44 As a lumped parameter mass balance model, the Burns model is incapable of accurately predicting the impact of on-farm management practices on water quality. But the SCS only needed a crude basinwide planning model to quantify the long-term effects of the Patterson Hollow Project on off-site water users and to justify the project to funding agencies.

Issue category	Drainage–specific issues	Institutional issues
	100 The Burns model might have helped the Southeastern Colorado Water Conservancy District in policy planning. But the model became a sensitive issue for the state of Colorado and was shelved, because Kansas might have benefited from the output of the model, which was in the public domain.	47 An individual farmer's on-farm management problems could be solved by going through various management options generated by existing computer programs. But the farmer rarely has the time or interest to sit down with the planner; he or she wants the problem solved immediately.
		99 Models have the potential for increasing our understanding of natural phenomena. But the water courts are constantly dealing with supposedly reputable models that none the less lead to conclusions that are 180 degrees apart.
		125 The SCS is looking at a relatively small-scale problem. But the Burns model was designed for basinwide scale, and has therefore a lot of simplifying assumptions that may not be appropriate at the SCS project scale.
		129 A model as a tool has the capability of decreasing conflict. But in the Colorado political context a model is going to increase conflict.
		171 For government people, the use of models puts things in a framework that is very understandable, and helps them justify what they are doing. But the local citizenry in the Arkansas Valley probably couldn't care less about models.
7 REGULATION	76 Colorado State Engineer has the authority to control waste of water, which could have a direct effect on water quality if it were strictly and vigorously enforced. But the State Engineer should be forthright about using diversion curtailments for water quality control, because water quality was not the underlying reason for granting the State Engineer the authority.	135 From the objectives of the Clean Water Act one could say that the EPA should be more involved in irrigation return flow policy. But since EPA does not get involved in matters of water quantity and instream flow, which this issue inevitably deals with, the agency does not have a strong role in the issue.

Table 4.2 (Continued)

Issue category	*Drainage–specific issues*	*Institutional issues*
	83 The State Engineer could probably control return flows through the authority to curtail waste of water. But an engineer telling a farmer his or her business would not be credible in a water court.	172 Colorado's Water Quality Control Division has an interest in irrigation return flow quality. But its hands are tied by the state's water laws, which prohibit regulating agricultural nonpoint sources in a way that would cause consumptive use of water or cause material injury to water rights, and direct regulators to use incentive, grant or cooperative programmes instead of regulations.
8 RESPONSIBLE PARTIES IN RETURN FLOW CONTROL	(No linkages)	68 The US Bureau of Reclamation operates the Pueblo Reservoir on the Arkansas River and is a wholesaler of water to the irrigators. But the Bureau does not really have a role in return flow policy, because it cannot control the application of irrigation water.
		169 The USDA probably has some cooperators among the Arkansas Valley farmers, who will participate in the Patterson Hollow project. But the area has mostly low-income farming, and it is unlikely that they will have the money to implement fully the changes necessary for an overall water quality improvement.

Notes
a Mutual irrigation company
b The Interactive Accounting Model (IAM)

more severe problems by threatening the water supply of the downstream irrigators (drainage-specific issues 38, 56 and 59 and institutional issues 42, 90 and 115). They point out that there is little incentive for farmers to improve their irrigation efficiency, because Colorado water courts have prohibited the sale of saved water (drainage-specific issue 22). The courts have not permitted a change in water rights if a transfer adversely affects a vested right of another appropriator in quantity, e.g. by diminishing return flow to downstream appropriators (Radosevich *et al.* 1976; Sutherland and Knapp 1988). Regulation is seen as another threat to irrigated agriculture, but the proactive stance the agricultural community has taken toward return flow management in projects such as the Patterson Hollow is postponing regulatory action (institutional issues 14 and 44). Water transfers may have had negative ecological and social impacts, but at least the water courts are beginning to realize this and have imposed conditions and restrictions on future transfers. And any legal changes that would restrict water transfers would in the experts' opinion cause even worse and unknown damage, because the right to freely buy and sell water is considered to be untouchable in Colorado (drainage-specific issues 48 and 65 and institutional issues 81, 110, 123 and 137).

In the end, the experts prefer the *status quo* to any changes in return flow management. According to them, irrigated agriculture under current irrigation practices may not last in the river basin in the long run, but any change in those practices would surely bring forth its downfall sooner. The interviewees cope with this dilemma by holding the cognitively dissonant position that while existing management practices should be dramatically altered to ensure long-term salt balance in the river basin, sticking to those same practices is none the less necessary as a way of ensuring the day-to-day managerial stability of the irrigation system. As a result, actions taken with the nominal objective of reducing salt load and improving water quality in the long run end up in fact being aimed at minimizing the operational cost and securing the short-term profitability of agricultural operations. Water quality is a concern only to the extent that local water quality control measures can fend off regulation by state or federal officials and secure funding for additional on-farm management projects.

The use of the IAM in the Patterson Hollow area reflects the wider institutional considerations of return flow management. The formal objective of modelling is to justify the local irrigation management project by showing that it has positive water quality impacts in the wider Arkansas River Basin (drainage-specific issue 52). In reality, the basin-wide model has little relevance to irrigation management modifications implemented in the field (institutional issue 125). The chief determinants of actions taken under the project are how successfully they cut an irrigator's operational cost and improve his productivity, neither of which is considered in the IAM. Rather than becoming an integral part of the planning of better on-farm management practices, the model takes on the role of a mediator between the project and the institutional support on which it depends. The model increases the chances of getting further funding for the local project (provided, of course, that it predicts positive water quality changes in the river basin); at the same time, however, the model becomes a cheap replacement for basin-wide monitoring of actual water quality changes (institutional issues 14, 44, 47 and 171).

The discrepancy between the stated and actual objectives of model application also creates a serious problem related to the geographical scale at which the IAM is used. As long as the functioning of the microscale (i.e. the on-farm systems included in the Patterson Hollow project) is unknown, the predictions obtained at the macroscale (i.e. the Arkansas River Basin) with the IAM are bound to be unreliable. Of course, political realities in the river basin work against identifying the sources and amounts of salinity. It is precisely the lack of interest by the SCS to pinpoint the sources of salinity in the Patterson Hollow project area and the emphasis on the economics of individual farming operations that have secured local support for the project.

Inadequate use of the IAM may seem sensible from the short-term political perspective. But inappropriate model application raises serious questions about the reliability of the basin-wide salinity impacts predicted by the model. Not only is the credibility of the IAM and the Patterson Hollow experiment put into question, but the future of any on-farm management project is ultimately jeopardized as well.

SCIENTIFIC ARGUMENTS IN POLITICAL CONTEXT

Analysis of the institutional context of modelling in the Arkansas River Basin indicates that inadequate model application and specification need not necessarily be the result of inadequate knowledge or experience in the art. What at a glance appeared as incomplete modelling reflected on closer inspection a profound, if paralysing, sensitivity toward the institutions and politics of water management in Colorado. Irrigation managers and water policy planners in other western states can learn from the principles of reform in the Arkansas River case, because the anomalies of western US drainage management have broadly similar institutional grounds. Two features in particular appear to be common to all arid states struggling with irrigation-induced water quality problems. The first has to do with the minor organizational role that drainage has historically played in overall watershed management. The second has to do with the role of scientific arguments, such as those based on modelling, in today's politically charged drainage controversies.

First, the 'drainage problem' is relatively new to western US water officials. The history of water development in the Arkansas River Basin and elsewhere in the arid American West underscores that irrigation, not drainage, has been the major concern of the large-scale water projects. There was little incentive to portray return flow as a problem, because it augmented the scarce water supply. Only in recent years have irrigation analysts worldwide begun to view drainage as a major subsystem of comprehensive water management instead of a low-priority activity within the operation and maintenance of a particular irrigation scheme. Drainage requires extensive planning, design, construction, operation and maintenance of facilities. Inadequate assignment of administrative responsibility for drainage has given rise to recommendations for organizational restructuring (Roe 1991).

In the Arkansas River case, the inappropriate application of the IAM reflects a fundamental organizational deficiency in the management of Colorado's agricultural return flows. Despite the crucial role properly handled drainage plays in securing the long-term viability of irrigated agriculture, no government organization in the state represents drainage interests. Instead, the administrative momentum is in securing the short-term profitability of agricultural operations. The absence of organizational support for return flow management has persuaded government officials and experts to operate as if irrigation were coming to an end, rather than strive for a sustainable agricultural economy. An autonomous drainage agency devoted solely to the task of constructing, managing and maintaining irrigation drainage in the basin would therefore have to be established to fill the organizational void.

Another characteristic of arid western US states is that efforts to find technical remedies to water management problems typically take place in a politically charged environment (National Research Council 1989). Legal and political water management disputes are particularly difficult to resolve, because opposing parties can find equally credible scientific support for their conflicting arguments. This, however, belongs to the nature of modern information societies and should not discourage technical experts from contributing their expertise to the resolution or better management of political debates. The minimum requirement for the debates should be that political arguments are scientifically reasonable. The sources of residual political disputes can then be traced to opposing value systems, differing premises of scientific theories, or conflicting policy interpretations allowed by scientific imprecision or inaccuracy.

In the Arkansas River case, not even the minimum requirement for a scientifically supported political debate existed. As the preceding analysis has shown, return flow experts themselves were uncertain about the characteristics of the region's geohydrology. There was some indication that the system had reached a dynamic long-term equilibrium of salinity, but conclusive evidence on the question did not exist. At the local level, experts had little detailed knowledge about the hydrologic interactions between the canal systems, the alluvial aquifer and the river. Without this knowledge, the sources and sinks of contaminants in the river basin and assessment of the viability of irrigated agriculture there will be subject to endless expert speculation.

A more troubling aspect of the Arkansas River case is that the scientifically enlightened debate, even with its minimum requirements fulfilled, would have taken place in a stifling institutional environment. The formal irrigation institutions in the state proved to be a systemic barrier against the provision of scientific support for effective return flow management. The institutional setting disabled the search for relevant drainage-related data, the utilization of that data as input for the right kind of return flow models, and the infusion of modelling results into the design of long-term return flow policies. I will elaborate the institutional reform implications of these issues in Chapter 8.

5

NETWORK ANALYSIS OF THE CONTROVERSY OVER IRRIGATION-INDUCED SALINITY AND TOXICITY IN CENTRAL CALIFORNIA

The Colorado case study in Chapter 4 treated expert interviews as the data of a qualitative expert system that can be used to better understand the institutional constraints to various technological interventions in the irrigation problem. The analytical approach in the second case study, which deals with the management of saline and toxic irrigation drainage in the San Joaquin Valley of central California, is the same, but the methods are more sophisticated. A combination of qualitative issue analysis and quantitative Bayesian network analysis reveals the mental constructs of key decision makers in the drainage controversy.

According to a fundamental premise of expert systems, expert statements can be viewed as components of a rule-based deduction method for solving problems (Nilsson 1980). To avoid incomplete deduction rules, expert system designers often interview several experts in the field. This, however, may exacerbate the problem of inconsistency if the interviewees use different conceptual models to elucidate the problem (Bramer 1985). Further difficulties emerge if the expert system is explicitly to take into account the attitudes (such as 'wanting' or 'disliking') the interviewees have toward the propositions incorporated in the rule-based deduction system (McCarthy 1985).

This chapter describes how the pitfalls of a traditional expert system can be avoided with Bayesian network analysis. Key experts and decision makers on the management of saline and toxic irrigation drainage in central California in the late 1980s had widely different and inconsistent causal explanations of the drainage problem. Furthermore, these explanations frequently included expert attitudes toward drainage management. Consequently, it was impossible to design a deduction method for solving the problem on the basis of individual expert rules. Bayesian analysis of the network of aggregated expert rules redefined the drainage problem. The inconsistencies and attitudes I found in individual experts' problem definitions are in the Bayesian approach direct indicators of potential remedies. According to the analysis, a systemic decision making dilemma persuaded drainage experts to avoid action toward a long-term, valley-wide drainage solution for the sake of immediate organizational and political benefits. Since the dilemma was embedded in the organizational and political context within which the experts operated, it was outside their immediate domain of knowledge and therefore escaped effective remedies. A prerequisite for successful drainage management in the valley is the restructuring of California's drainage authority.

In specifying the need for restructuring, this chapter expands the institutional proposition presented in Chapter 2. To recapitulate, this book as a whole elaborates the feedback between institutions and expert thinking, in which short-term oriented institutions advise environmental experts to hold the cognitively dissonant belief that short-term economics, however contrary to their convictions, must guide their decisions. The influence of formal institutions increases the dissonance between an expert's preferences and the institutional constraints, which in turn urges the expert to reduce that dissonance by acting in accordance with formal institutions. In this chapter I will show that the proposition applies on all three levels of an institution, including the operational rule level: the institutional problem is not only the incompatibility between short-term economics and long-term sustainability, or between different stages of environmental technology, but also between implementation and regulation of environmental technology.

The analysis has two parts. First, I will formulate the proposition of a drainage dilemma and the call for organizational restructuring on the basis of cognitive mapping of expert interviews. I will then verify the results of the interview analysis by investigating a number of case histories of drainage management, obtained from related literature of the time, in the context of a so-called sociotechnical matrix. This analysis illustrates the systemic spreading and penetration of the decision making dilemma into every aspect of agricultural drainage management in central California.

PHYSICAL AND SOCIAL CONTEXT OF IRRIGATED AGRICULTURE IN THE SAN JOAQUIN VALLEY

The San Joaquin Valley is the southern portion of California's Great Central Valley (see Figure 5.1). The 8.5 million acre (3.4 million hectare) valley floor was mostly wetland and semi-desert before small-scale irrigation by local irrigation districts during the first half of the twentieth century and subsequent large-scale water diversions from the Sierra Nevada streams by the federal and state water projects in the 1950s and 1960s (Letey *et al.* 1986; San Joaquin Valley Interagency Drainage Program 1979). Extensive reclamation and irrigation have turned the land into an agricultural region that supports annually a 15 billion dollar agricultural economy and produces a variety of sub-tropical crops ranging from citrus, avocados and kiwi fruit in the hotter southern regions to cotton, wheat and almonds in the more moderate northern areas (California Department of Water Resources 1987). In 1987, when this study began, six of the ten Californian counties with the highest total value of agricultural production were in the San Joaquin Valley, representing over half of the state's total value of agricultural production (Fay *et al.* 1991). But as in so many other areas of the world, intensive irrigation of arid or poorly permeable land has brought

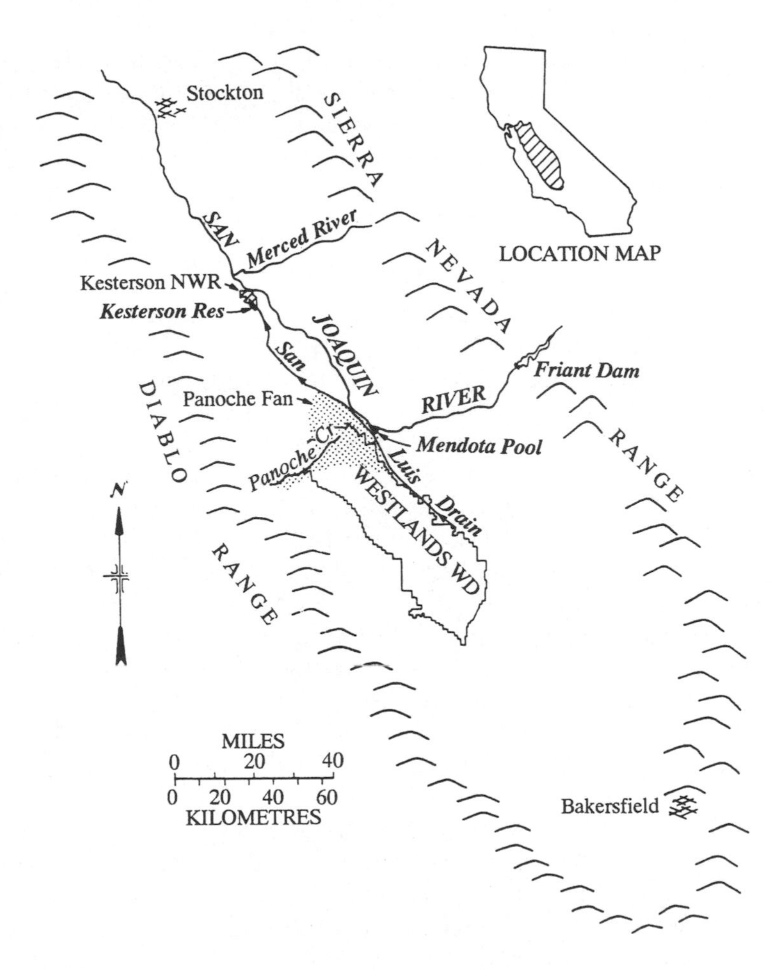

Figure 5.1 San Joaquin Valley, California (Moore 1989: 24).

with it drainage problems (Kovda 1983). Indeed, some 760,000 acres (310,000 hectares) of the valley's 4.7 million irrigated acres (1.9 million hectares) now have one or more of the following problems: waterlogging, salinization and toxicity (San Joaquin Valley Drainage Program 1989).

Waterlogging problems arise because of a poorly permeable clay layer that underlies the valley's primordial marine sediments at a depth of 400 to 850 feet (120 to 260 metres). Since groundwater seepage through the clay is very slow, intensive irrigation has raised the shallow groundwater level close to soil surface, depriving roots of oxygen and causing crop productivity to decline. Waterlogging problems are compounded by evaporative concentration of salts and trace elements in the soils and evaporation ponds of the arid western and southern valley. Salinity diminishes crop production by deteriorating soil properties and by causing toxicity to plants as a result of excessive concentration of boron, sodium and chloride ions. Toxicity of agricultural drainage to wildlife became evident in the early 1980s, when fish and water-bird deaths and deformities were discovered at the valley's Kesterson reservoir, which was formerly both a wildlife refuge and an evaporation pond for subsurface drainwater. The deaths and deformities were attributed to elevated concentrations of the trace element selenium in pond water and sediments. Selenium and other potentially toxic trace elements, such as molybdenum, arsenic, mercury and cadmium, occur naturally in the valley's sediment soils and are leached out by irrigation water (Grismer *et al.* 1988; Letey *et al.* 1986; Moore 1989; Ohlendorf 1986; San Joaquin Valley Drainage Program 1989; Tanji *et al.* 1986).

It has been estimated that by the year 2000, waterlogging problems – i.e. a groundwater table within 5 feet (1.5 metres) of land surface – will affect more than 900,000 acres (360,000 hectares) of irrigated land in the valley. More than 400,000 of those acres (160,000 hectares) will be underlain by water with quality problems, i.e. a salinity above 3,200 parts per million (ppm) total dissolved solids (TDS), and/or boron (B) above 8 ppm, and/or selenium (Se) above 5 parts per billion (ppb) (San Joaquin Valley Drainage Program 1989).

To resolve waterlogging and salinization problems, irrigation agencies have constructed subsurface drains to remove the drainwater from the fields. For more than forty years, a master drain to the San Francisco Bay–Delta area or a pipeline to the Pacific Ocean were considered to be the most feasible means of disposing of drainage out of the valley. As neither has been constructed, drainage has been discharged into the San Joaquin River or into evaporation ponds in the valley. Ponding in many parts of the valley, in turn, has introduced toxicity problems (Fujii 1988; Schroeder *et al.* 1988).

Valley drainage problems and their threat to irrigated agriculture have been recognized since the late nineteenth century. At that time, researchers emphasized the importance of ensuring salt balance in the soils under cultivation, but were less specific about the disposal of saline drainwater

(Elliott 1904; Fortier and Cone 1909; Hilgard 1886). Irrigation-related studies during the first half of the twentieth century frequently made note of waterlogging and salinization problems. Efforts to achieve salt balance were only partially successful, however, and were eventually overridden by the greater concern to expand the state's large-scale water projects in the 1930s (California Department of Public Works 1930, 1931; Kelley and Nye 1984). Organized governmental efforts aimed specifically at eliminating the problems did not begin until the 1950s. Three periods of organized drainage-related research can be distinguished (Hukkinen 1990; Hukkinen *et al.* 1990).

Between 1950 and 1975, the dominant theme in federal and state drainage research was to determine the technical and economic feasibility of constructing a master drain to convey drainwater from the entire valley to the Sacramento–San Joaquin Delta in the north. The master drain was not built because of local irrigators' reluctance to finance the drain and because of environmental concerns expressed by San Francisco Bay area interests. Instead, the US Bureau of Reclamation (USBR) constructed a shorter version of the drain, the San Luis Drain, which led drainwater from the northern Westlands Water District to the Kesterson reservoir (Figure 5.1) (California Department of Water Resources 1957, 1965, 1974; California Senate Permanent Fact Finding Committee on Water Resources 1965; San Joaquin Valley Drainage Advisory Group 1969; US Bureau of Reclamation 1945, 1964).

From about 1975 to 1982, federal and state studies began to compare and evaluate several alternative drainage management options besides the master drain, such as disposal to the Pacific Ocean via a pipeline, evaporation ponds, desalination and reuse, and discharge to the San Joaquin River. The scope of research broadened to include not just consideration of technical and economic feasibility but also assessment of environmental impact of the proposed remedies. In the end, the master drain concept was always found to be the best solution. A detailed benefit–cost analysis comparing a number of long- and short-term alternatives for drainage management found the net disposal benefit of a master drain to the San Francisco Bay–Delta region to be 18 million US dollars, compared to a net benefit of 2 million US dollars for making no long-term, valley-wide drainage investments (San Joaquin Valley Interagency Drainage Program 1979).

The discovery of bird and fish kills at Kesterson reservoir in 1982 and the subsequent disappearance of the master drain concept as a drainage remedy mark the beginning of the final research phase. In 1984 federal and state agencies launched a joint task force, the San Joaquin Valley Drainage Program (SJVDP), to resolve the long-term, valley-wide drainage issue. After the press accused SJVDP of promoting the disposal of toxic drainwater to the San Francisco Bay area and the Pacific Ocean, the Program's management directed its staff to consider only in-valley and short-term solutions to the problem. In addition to technical, economic and environmental feasibility studies, SJVDP funded research assessing the risk, social and institutional impact, and political and legal feasibility of numerous in-valley remedies. Despite optimistic anticipation at the outset of the seven-year research effort to 'develop comprehensive plans for long-term management of [drainage] problems' (San Joaquin Valley Drainage Program 1987: 2) the Program's final report admits that 'no single, sure, and lasting solution to the drainage problem has been put forward' (San Joaquin Valley Drainage Program 1990: iii).

Political considerations have dominated the drainage issue since the discovery of toxic compounds in drainwater. Public awareness of the risks of selenium and other potential toxic trace elements has increased the conflict and uncertainty over disposal. Attention has therefore shifted to evaluating the short-term alternatives, such as improved on-farm management practices and various biological and physical–chemical treatment processes. The examples presented below detail how and why the crucial question of disposal has been abandoned in favour of the short-term relief promised by on-farm management and treatment.

Combined with the decades-old criticism against agricultural crop and water subsidies, the agricultural and environmental problem of drainage has turned into a sociopolitical issue that threatens to undermine the legitimacy of California's irrigated agriculture (Hukkinen 1990; Hukkinen *et al.* 1990; Taylor 1979). The critics of agribusiness can find any number of economically or environmentally more beneficial uses for the water that currently goes to agriculture, including urban use in the state's rapidly growing cities, and in-stream use to accommodate fish habitat and human recreation (California Department of Water Resources 1987; Reisner 1989).

NETWORKED PROBLEMS AND CONSTRAINED SOLUTIONS

The failure of past research to lead to implemented remedies suggests that the valley's agricultural drainage 'problem' has other dimensions besides the much-studied technical, economic and ecological; and that the 'solution' may involve more than finding the most feasible one among alternative technical remedies. Twenty-three interviews conducted in 1987 with drainage experts in California's irrigation and regulatory agencies and environmental groups corroborate this conclusion.

The interviewees were categorized into four groups: twelve were associated with the agricultural community (local water, irrigation or drainage district managers); six were classified as planners (representatives of the SJVDP, California Department of Water Resources [DWR], and California Department of Fish and Game); three were regulators (staff from the State Water Resources Control Board [SWRCB], Central Valley Regional Water

Quality Control Board, and the US Environmental Protection Agency [EPA]); and two were from the environmental community (the Natural Resources Defense Council and the San Francisco Bay Institute). The agricultural community and planners are together called California's 'irrigation bureaucracy', i.e. operational and research staff associated with federal and state irrigation agencies and local districts. Each interest group was and still is an important party to the ongoing drainage debate: the agricultural community represents local private and public interests of the irrigation bureaucracy; planners are experts from the centralized public sector of that bureaucracy; regulators set crucial environmental constraints on irrigation agencies' operations; and the environmental community has in the past few years become one of the most influential critics of California's irrigated agriculture (Johns and Watkins 1989; Kahrl 1979; Letey *et al.* 1986; Reisner 1987; Worster 1985).

During the loosely structured interviews, I asked the drainage experts to describe the San Joaquin Valley's agricultural drainage problems and possibilities for solving them. A comparison of the perceived problems and solutions revealed a viciously systemic impasse, in which short-term organizational and political imperatives led key decision makers to tolerate a situation untenable from the point of view of long-term resource management.

Bayesian problem networks

As described in Chapter 3 (pp. 38–9), the interviewees typically told conflicting narratives, which included both empirical and normative claims about the factors perceived to be contributing to current drainage problems. No set of problems could therefore be singled out as the primary target for remedial action. But the problem statements that the experts made in their interviews were frequently causally linked, which indicates that the opinions might be aggregated into an expert system. Expert systems accept expert statements as building blocks of a rule-based deduction method for solving the problem at hand (Nilsson 1980). While the problem statements and their causal linkages do not specify rules for solving the drainage problem, they none the less provide an understanding of its underlying cause-and-effect linkages. As such, they serve as indicators of potential remedies.

The problem statements and links between them were used as the building blocks of a problem network and analysed as detailed in Chapter 3 (pp. 41–4). Among the most prominent characteristics of the problem networks were loops, which are of much interest to policy planning. They reflect circular argumentation in an issue and blur the crucial distinction between cause and effect. Loops also provided an effective criterion for determining the appropriate level of problem network aggregation in the

Table 5.1 Emergence of loops in paired aggregation of interest group networks in the San Joaquin Valley drainage debate (number of loops)

	Agricultural community	*Planners*	*Regulators*	*Environmental community*
Agricultural community	3	5	16	4
Planners	5	0	11	0
Regulators	16	11	3	3
Environmental community	4	0	3	0

Source: Hukkinen 1993a: 187.

San Joaquin Valley case study. When networks were aggregated at the level of interest groups (by constructing four aggregated networks, one each for agriculturalists, for planners, for regulators and for environmentalists), only six loops appeared (Table 5.1). This indicates that the four interest groups were in fact fairly distinct and homogeneous, since people or groups rarely justify their actions knowingly in terms of circular argumentation (Simon 1964). (However, as will be seen in Chapter 6, when speculating about environmental problems of the very distant future, individual experts appear less averse to arguing in circles.) In contrast, paired aggregation of group networks produced thirty-nine loops. This indicates the existence of critical interfaces between interest groups: experts from two different interest groups share problem statements but link them in a causally inconsistent fashion.

The six problem networks obtained in paired aggregation of the four interest group networks still contain too many problem statements to be feasibly analysed. A method is therefore needed to estimate the relative importance of the component problems and the causalities that link them. The frequency of mention of problem nodes and links in network aggregation provides a crude measure of the relative importance, or probability, of different parts of the network. But the method of analysis must also take into account the causal dependencies a problem statement has with the surrounding problem network. The Bayesian probability concept does precisely this (Duda *et al.* 1976; Kim and Pearl 1983; Pearl 1986, 1988).

Conditional relationships between probabilistic events are the basic building blocks of Bayesian probability theory. The famous Bayesian inversion formula,

$$P(H|e) = [P(e|H)P(H)]/[P(e)] \quad (1)$$

states that the belief we have in hypothesis H after obtaining evidence e can be computed by multiplying our previous belief $P(H)$ by the likelihood $P(e|H)$ that e will occur if H is true. $P(H|e)$ is the posterior probability, $P(H)$ the prior probability, and $P(e)$ a normalizing constant, $P(e) = P(e|H)P(H) + P(e|\neg H)P(\neg H)$, which makes $P(H|e)$ and $P(\neg H|e)$ sum to unity (note that '¬H' stands for 'not-H') (Pearl 1988; Press 1989; Schmitt 1969).

Equation (1) is significant because it expresses a probability $P(H|e)$, which people often find hard to assess, in terms of quantities that are usually known from experience. Imagine you are in a casino and hear the person at the next gambling table declare the outcome 'Twelve!' You wish to know whether he was rolling a pair of dice or spinning a roulette wheel. Your knowledge of gambling devices yields the quantities $P(Twelve|Dice)$ and $P(Twelve|Roulette)$ – 1/36 for the former and 1/38 for the latter. You can also obtain the prior probabilities $P(Dice)$ and $P(Roulette)$ by estimating the number of roulette wheels and dice tables at the casino. Plugging this information into the Bayesian inversion formula yields the posterior probability you were looking for, $P(Dice|Twelve)$. Issuing a direct judgement $P(Dice|Twelve)$ would have been much more difficult – something only an expert in that particular casino might have done reliably (Pearl 1988).

The Bayesian conditional probability $P(e|H)$ conveys the degree of confidence in rules such as 'If H then e', and is therefore a convenient way of expressing the magnitude of causal dependency. A simple tree-like Bayesian network representing the probability distribution $P(x_1,x_2,x_3) = P(x_3|x_1)P(x_2|x_1)P(x_1)$ is presented in Figure 5.2. The nodes x_1, x_2, and x_3 represent variables, the links $x_1 \rightarrow x_2$ and $x_1 \rightarrow x_3$ signify the existence of direct causal influences between the connected variables, and the strengths of these influences are expressed by forward conditional probabilities $P(x_3|x_1)$ and $P(x_2|x_1)$ (Pearl 1988). The only parameters needed for the construction of a Bayesian problem network are estimates of the conditional link probabilities. In some expert systems, the estimates are made by individual experts (Forsyth 1984; Negoita 1985; Pearl 1988; Weiss and Kulikowski 1984). In the San Joaquin Valley drainage case study, drainage experts collectively make the estimates. For example, if the link $X_1 \rightarrow X_2$ in Figure 5.2 were mentioned by three interviewees, and the link $X_1 \rightarrow X_3$ by two interviewees, the conditional link probabilities would become $P(X_2|X_1) = 3/(3+2) = 0.6$ and $P(X_3|X_1) = 2/(3+2) = 0.4$. In other words, the arrows exiting a given problem node describe completely what the interviewees perceive to be the possible effects of that particular problem statement, whereby the sum of the conditional link probabilities must be unity.

A Bayesian network is capable of representing the generic knowledge of a domain expert – in this case the knowledge of a drainage management expert. Two different types of probabilities characterize the network. Fixed conditional probabilities label the links, whereas the nodes store the dynamic values of the updated node probabilities, or beliefs (BEL).

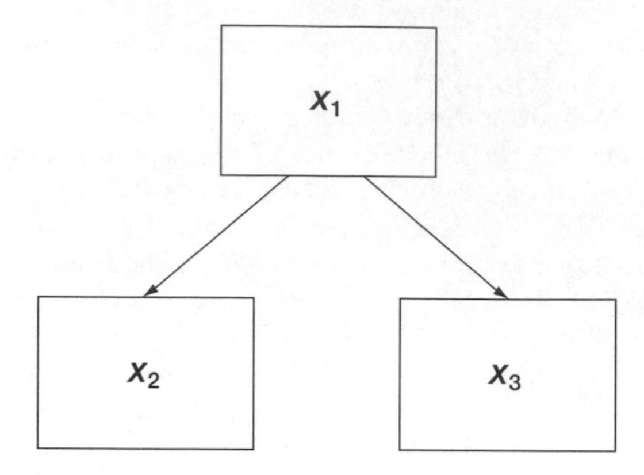

Figure 5.2 A tree-like Bayesian network representing the distribution $P(x_1,x_2,x_3) = P(x_3|x_1)P(x_2|x_1)P(x_1)$.

Computations in the network involve fusing and propagating the impacts of beliefs through the network until equilibrium is reached (Pearl 1986). At equilibrium, each problem statement has been assigned a certainty measure consistent with probability theory. Under the computational scheme it is possible to determine the experts' collective strength of belief in a problem statement, given their strength of belief in the rest of the problem network surrounding the problem statement.

The network in Figure 5.2 has a simple, tree-like structure. In reality, aggregated problem networks have the appearance of entangled fish-nets, with several open-ended problem paths (beginning with initial problems or ending in terminal ones) at the edges and a web of complex branching loops in the centre. The mathematical treatment of such complexities is summarized in Appendix 2. Suffice it to say that the loops in problem networks are positive feedback loops in which problems reinforce themselves. In probabilistic terms, such a network configuration increases the likelihood of the loop problems persisting indefinitely, making their probabilities approach unity.

Constraints on solutions

The experts imposed constraints on potential drainage solutions which reflected the logical inconsistencies found in problem definitions. These constraints were expressed in exactly the same format as were the issues in the Colorado case study: the interviewees stated a potential solution, only to rebut it with a cognitively dissonant counter-argument. But in contrast to problem definitions, the experts very much agreed on the limitations to future drainage remedies. Three interrelated issues, or operating assumptions, could be identified in the interviews:

1 The no-drain assumption. No system-wide master drain to the Delta is possible at this time or in the foreseeable future, but it is none the less important to manage the valley's agricultural drainage effectively (twenty-one of the twenty-three interviewees stated this assumption).
2 The no-regulation assumption. Specific drainage-related regulation is unlikely to succeed, but it is none the less important to enforce valley-wide regulations as part of the overall solution (seventeen interviewees agreed on this).
3 The no-responsibility assumption. Specific regional cooperation in drainage is impossible in the valley, but collective responsibility is none the less necessary for effective drainage management (seventeen interviewees agreed on this).

Comparison of the collective problem definitions and boundary conditions for solving the perceived problems enables a redefinition of the San Joaquin Valley's drainage problem at a systemic level. The new problem conception is systemic, because it incorporates not just the technical, economic and ecological, but also the organizational and political aspects of the issue. Furthermore, analysis of the experts' collective definitions of problems and solutions includes both their political values and what they perceive to be scientific facts. This is crucial, because the success of conflict resolution in a politically charged situation should and often does depend on society's ability to create acceptable political decision making mechanisms that take into account the whole spectrum of ideological principles and attitudes held by the interested parties (Nelkin 1984).

With these assumptions, analytical tools and network modifications it is possible to present the new insights that the Bayesian network approach has to offer to the San Joaquin Valley's agricultural drainage problem.

ANALYSIS OF AGGREGATED NETWORKS

Bayesian analysis reduced the six aggregated drainage problem networks into highly probable networks with two distinct structural features: loops and open-ended networks surrounding the loops. The network that emerged when the networks of agricultural and planning experts (i.e. the irrigation bureaucracy) were aggregated exemplifies the five other aggregated networks, and will therefore be described in detail (for a detailed analysis of all six networks, see Hukkinen 1990).

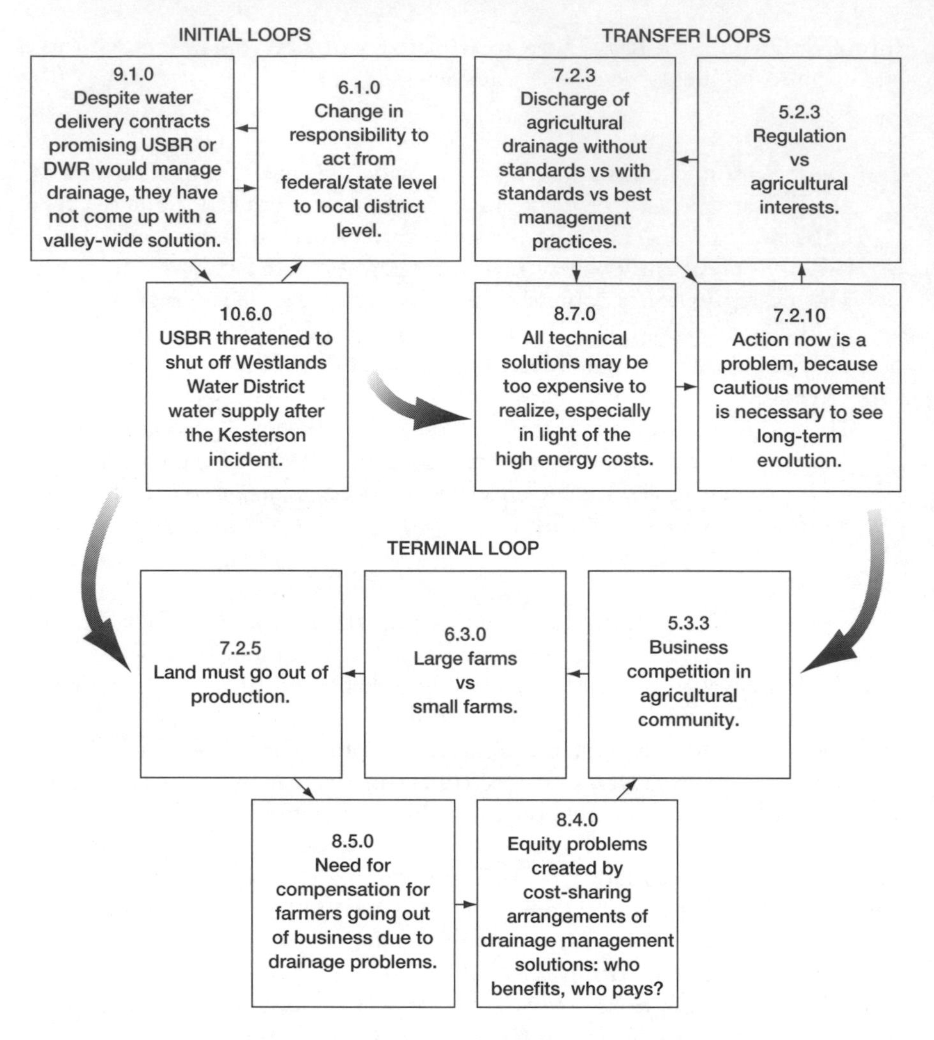

Figure 5.3 Loops of the agricultural community and planners, i.e. the irrigation bureaucracy, in the San Joaquin Valley drainage debate.

Irrigation bureaucracy's loops

The loops of California's irrigation bureaucracy were in three clusters: two initial loops led both directly and via two transfer loops to one terminal loop (Figure 5.3). According to the first initial loop, polarization of the drainage debate was possible because of mutual reinforcement of two phenomena: the agricultural community's conviction that the central planning agencies were still responsible for valley-wide drainage management, despite their failure to build the master drain (problem statement 9.1.0); and the tendency among planners to place the responsibility for drainage on local districts (6.1.0). The accusations in the first initial loop were amplified by the USBR's threat to shut off Westlands Water District's irrigation water supply immediately after the discovery of dead and deformed water-birds at the valley's Kesterson reservoir (10.6.0).

The initial loops' problems led to more uncertainty and polarization in the two transfer loops. Basic conflicts of interest between the agricultural community and regulators caused disputes over standards for water quality (7.2.3). Because all known technical solutions to the drainage problem might in the end be too expensive (8.7.0), interviewees felt slow and cautious action was necessary from the outset (7.2.10). Yet caution and delay led back to and further increased the conflicts of interest between agriculturalists and regulators (5.2.3), particularly since regulators were under pressure to deal with the issue of standards rapidly and decisively.

Finally, both the initial and transfer loops led to the terminal loop. Agricultural experts said that the necessity of taking land out of irrigated production (because of severe drainage problems) raised the need to compensate the affected farmers (7.2.5 and 8.5.0). This posed questions about the equity of compensating the affected farmers for production losses and not other farmers who also suffered from drainage-related problems but who remained in irrigated production (8.4.0). In planners' opinion the agricultural community's recognition that drainage costs were rising for all farmers but that government compensation was not, only induced further competition among farmers for what remaining profits were to be made from irrigation (5.3.3). Increased competition in turn set large farms against small ones (6.3.0), inevitably forcing farms with severe drainage problems out of irrigated production (thus closing the loop in statement 7.2.5).

The substance of the problems in the irrigation bureaucracy's loops revolves around five partially overlapping issues: (1) uncertainty in drainage management; (2) performance of drainage solutions; (3) performance of drainage regulation; (4) responsible party in drainage; and (5) polarization of drainage issues. Irrigation bureaucracy's initial loops, for example, deal with questions of uncertainty (problem statements 6.1.0 and 10.6.0), the performance of drainage solutions (9.1.0 and 10.6.0), responsibility in drainage management (6.1.0 and 9.1.0), and polarization (9.1.0 and 10.6.0). The transfer loops contain issues of uncertainty (7.2.3 and 8.7.0), the performance of drainage solutions (7.2.10 and 8.7.0), the performance of regulation (5.2.3 and 7.2.3), and polarization (5.2.3, 7.2.3 and 7.2.10). Finally, the terminal loop includes matters concerning the performance of drainage solutions (7.2.5 and 8.5.0) and polarization (5.3.3, 6.3.0 and 8.4.0) (Figure 5.3). These issues also govern the networks surrounding the irrigation bureaucracy's loops.

Networks surrounding the loops

The networks surrounding the loops contained too many problems for the purposes of a detailed content analysis. Criteria were needed for determining the cut-off point, i.e. the boundary between problems to be included in further analysis and those to be excluded from it. Two criteria determined the cut-off point. First, the number of problem statements included should be comparable to that found in the loops (approximately ten). Second, the cut-off point should coincide with a significant drop in belief when the problem statements are ranked in the order of diminishing belief.

The substantial issues raised in the loops also emerged in the highly

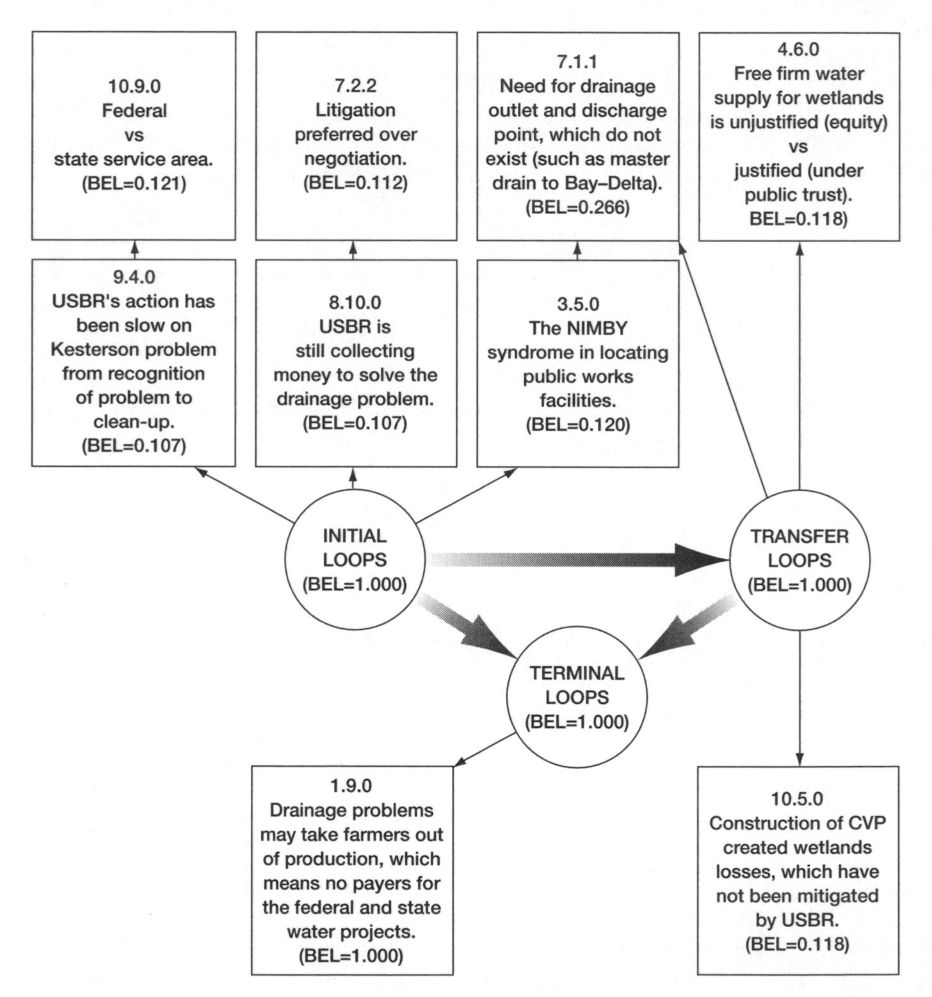

Figure 5.4 Highly probable network surrounding the loops of the irrigation bureaucracy in the San Joaquin Valley drainage debate.

probable network surrounding the loops, but the predominant structure was a path of problems leading to a terminal problem (Figure 5.4). The two terminal problems in which agricultural and planning experts believed the most were: 'Agricultural drainage problems may take farmers out of production, which means no payers for the federal and state water projects' (problem statement 1.9.0); and 'Need for an agricultural drainage outlet and discharge point, which do not exist (such as the master drain to the Bay–Delta)' (7.1.1). These problems succinctly summarize the ultimate concerns of the irrigation bureaucracy in the drainage debate. First, uncertainties concerning the responsible party in drainage management (initial loop in Figure 5.3) and the possible non-feasibility of any technical solution to the drainage problem (transfer loop in Figure 5.3) were perceived to be leading to the ultimate impasse where no relief to drainage problems could be expected before the notoriously absent master drain was in place. Second, increasing land retirement (as described in the terminal loop) might ultimately threaten the very lifeline of irrigated agriculture, namely its water supply. In other words, drainage problems were seen to have the potential not only for resisting any solution but also for threatening the survival of irrigation bureaucracy by questioning its autonomy, authority and influence.

Even to raise the question of land retirement was correctly perceived by irrigation experts as a battle call. Heated interviewee comments indicate that land retirement was an incomprehensible proposal for many San Joaquin Valley irrigators in the late 1980s: 'Land will NOT go out of production.' On the other hand, land retirement was no problem at all for segments of the environmental community – indeed, it was the solution. 'Concretely [the drainage problem] means that some land will have to come out of production,' as one environmentalist argued.

The rest of the irrigation bureaucracy's problems surrounding the loops described other pathways leading to both intra- and inter-organizational conflicts. These conflicts were over drainage responsibility between federal and state service areas (10.9.0), over wetlands losses created by irrigation projects (10.5.0), and over the water supply that wetland areas were allegedly entitled to under California's public trust doctrine (4.6.0). In general, the experts noted the polarizing effects of the tendency to litigate rather than negotiate in Californian water issues (7.2.2) (Figure 5.4).

A discouraging picture emerges when the network in which the irrigation bureaucracy probably has a high level of belief is considered as a whole. Uncertainty over drainage solutions, regulation and responsibility is seemingly intractable, because no causal origin can be pinpointed to the problems at hand. Instead, the experts' circular arguments increase issue uncertainty and the potential for organizational conflict. Such conflict is further polarized by the key beliefs that the lack of the desperately needed master drain (7.1.1) may signal the downfall of irrigated agriculture (1.9.0).

The role of cognitive dissonance

Now compare the findings of network analysis with the operating assumptions that the experts perceived to be constraining potential drainage remedies. Irrigation experts' strongly expressed belief that there should be a master drain (7.1.1) stands in sharp contrast with their operating assumption that there will be no system-wide master drain now or in the foreseeable future. Cognitively dissonant decision makers such as these typically have ambiguous and uncertain goals (March and Simon 1994). In our case, the uncertainty ripples throughout the organizational hierarchies involved in the San Joaquin Valley drainage controversy, from individual experts to entire interest groups (Hukkinen 1991c). The irrigation bureaucracy has acquiesced to this highly unsatisfactory situation, because the payoff of living with uncertainty has been the avoidance or postponement of organizational conflict. What or whom could the environmentalists argue against, when irrigators openly say that no drain to the Delta will be constructed?

Another preference implied in the irrigation bureaucracy's problem network is that some party to the drainage debate should assume chief responsibility for drainage management (problem statement 3.5.0 in Figure 5.4). This preference is in contrast to the widely shared operating assumption that no interest group is likely to cooperate or assume responsibility in drainage management. Avoiding responsibility has prolonged the physical problems of waterlogging, salinity and toxicity, but has at the same time prevented a polarized political debate over specific drainage remedies, such as the master drain or toxic treatment technologies.

Similar dynamics could be found in the problem networks to which regulators contributed. Once again, uncertainties concerning drainage regulation, solution and responsibility were presented in the form of circular arguments that led to terminal problems describing the potential for various organizational conflicts. The preference regulators expressed in these networks was that some type of drainage regulation should take place, despite difficulties encountered in enforcement. This, of course, conflicts with the operating assumption that specific regulations will not succeed in the valley. By avoiding stringent regulations, state regulators insulated themselves from the criticism of the state's politically influential agribusiness interests. Cognitive dissonance over drainage responsibility and the master drain was also reproduced in the networks constructed between regulators and the irrigation bureaucracy. The shared cognitive dissonances of irrigators and regulators ensured a virtual paralysis of the valley's long-term drainage management.

The rest of the aggregated problem networks reflected the organizational and political anomalies in the relationships between the irrigation bureaucracy, regulators and the environmental community. Substantially, the component problem statements could be classified into the same five categories that characterized the irrigation bureaucracy's networks, namely, uncertainty in drainage management, performance of drainage solutions, performance of drainage regulation, responsible party in drainage, and polarization of drainage issues. Problems in the last category, i.e. those describing various kinds of organizational and political conflict, increased dramatically in the networks surrounding the loops. Such problems represented 65 per cent of loop problems and no less than 91 per cent of problems surrounding the loops. Structurally, five of the six aggregated networks were similar, characterized by a set of loops surrounded by terminal problem paths (the aggregated network of planners and environmentalists had no loops, but was otherwise similar to the other five networks). Furthermore, a detailed morphological analysis of the loop structures found that the twenty-seven different loops in the networks could be reduced to seven relatively simple 'fundamental loops' (Hukkinen 1990). The irrigation bureaucracy's loops in Figure 5.3 are representative of the other loops, as they contain elements of each fundamental loop.

In sum, Bayesian analysis shows that irrigators and regulators faced the need to reduce uncertainty about drainage treatment and disposal, but risked polarizing the irrigation controversy with any reduction in this uncertainty. This systemic drainage dilemma prevented action-forcing decisions about the management and regulation of the San Joaquin Valley's agricultural drainage.

THE DRAINAGE DILEMMA

The interviews indicate that, in the aftermath of the Kesterson incident, Californian irrigators and regulators were confronting a fundamental conflict between what they thought should be done about drainage problems and what they considered the feasible course of action. From the viewpoint of the irrigation bureaucracy, politics after Kesterson required that agricultural drainage be guided primarily by the ideals of long-term environmental protection. Yet the bureaucracy as a whole continued to stress the short-term economics of agricultural production. Any effort by the irrigation bureaucracy to manage drainage, most of all by constructing the master drain, would have secured agricultural production over the long haul. But, at the same time, it would have raised criticism from the environmental community and the state's urban electorate against toxics management and irrigated agriculture. Not managing drainage might have quelled such criticism in the interim, but eventually it would have led to further agricultural *and* environmental difficulties in the form of extensive land retirement and toxicity problems. Thus, every decision bureaucrats made about drainage was characterized by ineffective tiptoeing between doing what looked like drainage management and not doing what was drainage management.

Now consider the regulators. They had to enforce stringent water quality and quantity regulations, if simply to convince California's environmentally conscious public and press that drainage toxics were not a hazard. At the same time, they had to avoid over-regulating the farmers, because expensive drainage management might have bankrupted irrigated agriculture. The regulators dealt with their decision making dilemma just as the irrigators did, that is, by accepting the view that, even though real regulation should in theory take place, it would not do so in practice. And just as the irrigation bureaucrats, the regulators could micromanage drainage with endless research projects and public hearings, the preoccupations of both SJVDP and SWRCB throughout the 1980s.

To disguise their uneasy walk between appearing to implement and actually failing to implement solutions, the state's irrigation bureaucrats and regulators tried to assign the responsibility to each other. Whenever the irrigation bureaucrats were accused of not being committed to resolving the problem, they asserted that the responsibility lay with the regulators and vowed to take action the moment environmental regulations were in place. Whenever the regulators were similarly accused, they reminded the critics that the prerequisite for successful drainage management was the local irrigators' cooperation in resource management. Passing the buck therefore established an alliance of mutual tolerance between irrigators and regulators. Each party thought the other party should take the decisive steps toward a comprehensive drainage management solution. But neither party publicly stated its preferences for fear of polarizing the debate to the extent that environmentalists and the public would question the legitimacy of irrigation.

By tolerating the conflicting goals of drainage management, irrigation bureaucrats and state regulators for the time being avoided having to defend themselves against those who claimed that irrigating California's semi-arid lands was a waste of precious water. Political opposition to the master drain and the ocean pipeline, voter rejection of new large-scale water projects in the state, efforts to restrict any new diversions from the San Francisco Bay–Delta system, and the controversy over the Kesterson clean-up were all reminders of how 'solutions' to water use problems quickly come under attack in California (California State Water Resources Control Board 1988; Engelbert and Munro 1982; Hart 1984; Horne 1988; Letey *et al.* 1986). When irrigation bureaucrats and regulators refused to move to any action-forcing direction while continuing to order further studies, their critics had few credible arguments to make against irrigation or drainage. Unfortunately, the bureaucrats' conflicting intentions also paralysed comprehensive, long-term drainage management. In sum, the contradictory goals prevented a polarized drainage conflict by confusing the political debate and by hindering action-forcing decision making. Both of these, however, only increased the same uncertainty that threatened to polarize the conflict in the first place.

This systemic dilemma thoroughly governed the policy makers' consideration of drainage management. It prevented any proposed solution from being implemented the moment research and development had reduced uncertainty to a level that would otherwise have permitted deployment. At the same time, recognizing the dilemma points out a fruitful avenue toward remedy.

Primarily organizational structures and incentives prevented irrigators from implementing drainage solutions, and regulators from establishing and enforcing regulations. The network analysis shows that, while mention is made both of irrigation-related factors and of drainage-specific problems in the loops, the sole terminal problem of the terminal loop in Figure 5.4 concerns the threat that drainage problems in particular pose to the long-term survival of the state and federal water projects in California. Nothing is implied with respect to irrigation *per se*. Yet at the time of the interviews drainage was commonly taken to threaten irrigation, because the former was assumed to be an inseparable physical and organizational component of the latter. While the physical relationship is inherent, the organizational one is not. Even though irrigation often does lead to salinity and toxicity problems, irrigation agencies not only can, but traditionally have, functioned without considering drainage. Maybe drainage problems would become more tractable if assigned to a specific agency not charged with the future of irrigation?

The source of the drainage dilemma lies in the mismatch between the preferences of irrigators and regulators, and the norms of influential critics in the surrounding society (such as environmental groups, politicians and the press). Irrigators and regulators want farming and ranching to flourish, whereas their critics question the legitimacy of government subsidized irrigated agriculture as a whole. As the environmental norms are critical toward irrigation but not toward drainage as such, return flow management needs to be organizationally separated from irrigation. In practice, this calls for an administrative arrangement similar to that proposed in the Colorado case study in Chapter 4 (see pp. 70–2), namely, the creation of an independent drainage bureaucracy with the sole responsibility for environmentally sound drainage management. Irrigation agencies would have to be relieved of drainage responsibilities. The arrangement would obviously not make the conflict between short-term economic profitability and long-term environmental sustainability disappear – it would just transform it from a cognitive into an inter-organizational conflict, which would have a better chance of being debated and decided upon openly. I will examine the theoretical foundations and practical details of this proposal in Chapter 8.

VERIFICATION OF THE DILEMMA: FRUSTRATED DRAINAGE MANAGEMENT

The proposition that Californian irrigation officials are stuck in a dilemma between uncertainty and polarization is based on interviews with decision makers and experts. But was there any evidence in the real-life drainage policy and management of the late 1980s California that such a dilemma existed? To gauge the conflict and uncertainty in the drainage controversy, I surveyed research reports, periodical articles and newspaper writings published on the issue after the Kesterson incident. After this initial survey, I

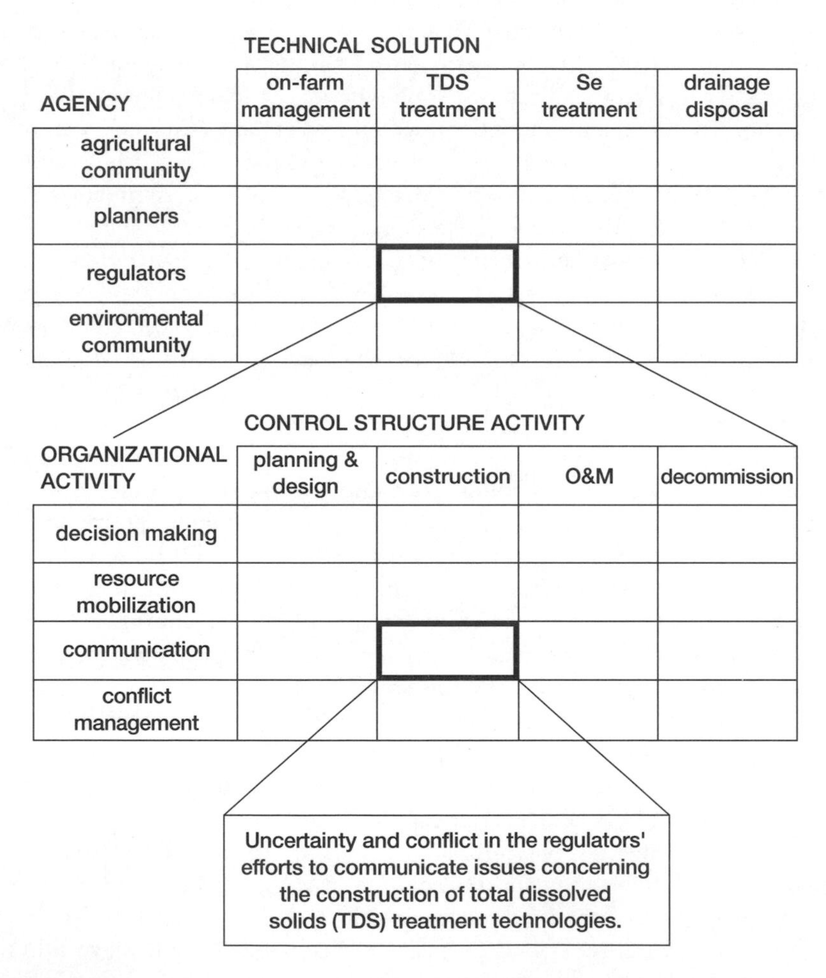

Figure 5.5 Socio-technical matrix for identifying potential sources of conflict and uncertainty in the San Joaquin Valley drainage issue.

chose a few case histories of drainage management efforts undertaken in the San Joaquin Valley during the 1980s for closer examination.

A systematic investigation of the complexities of drainage management calls for a framework that takes into account both technological and organizational dimensions. Figure 5.5 presents one possible framework, the sociotechnical matrix. It is a modified version of a framework developed by Uphoff (1986) for assessing irrigation bureaucracies and technologies. In the sociotechnical matrix, each proposed technical solution to the drainage problem is examined in terms of four control structure activities: planning and design; construction; operation and maintenance (O&M); and decommissioning. The agencies and parties undertaking these tasks are examined in terms of four organizational activities: decision making; resource mobilization; communication; and conflict management.

Technical solutions considered here include on-farm management, total dissolved solids (TDS) treatment, selenium (Se) treatment and drainage disposal. Agencies and parties are again grouped into agricultural community, planners, regulators and environmental community.

Control structure activities and organizational activities are subcategories of technical solutions and agencies, respectively. Yet the interactions between these sets of activities are of most interest for the current analysis. Consequently, the sociotechnical matrix is designed to facilitate the explication of linkages among the physical activities that control drainwater flow and the organizational activities that take place in the relevant agencies. For example, planning and design require not only decision making, but also resource mobilization, communication and conflict management within and among agencies. Construction involves decision making about how specific tasks are carried out, substantial resource mobilization, and much communication and conflict management as the work is being done. O&M call for decision making about schedules and work assignments, mobilization of resources such as information, labour and funds, and management of conflict over what requires O&M, how it should be done, and by whom (Uphoff 1986). Finally, as the following examples illustrate, the decommissioning of drainage facilities involves complex processes of decision making, communication, conflict management and resource mobilization.

These cross-cutting categories provide the framework for identifying and discussing the uncertainties and conflicts that have arisen in each examined case history when agencies and parties have interacted with potential solutions. I will first identify the agencies responsible for implementing a technical solution and then discuss the relevant elements in the sociotechnical matrix. Not all possible permutations in the sociotechnical matrix are of interest to the present analysis, particularly since most of the proposed technical solutions were still in the planning and design stage at the time of the analysis, and the actors in the drainage debate were not equally involved in the various organizational efforts to deal with the controversy. Thus, the following sections will focus on the most important structural and organizational activities.

On-farm agricultural and irrigation management

The proposed on-farm solutions to deal with the drainage problem can be divided into two options, namely land retirement and best management practices (BMPs) (Grismer *et al.* 1988; Hanemann *et al.* 1987; Letey *et al.* 1986; San Joaquin Valley Drainage Program 1987). From the point of view of irrigators, better on-farm management is more 'positive' than land retirement, in the sense that management practices work toward keeping or increasing the viability of irrigated agriculture. There is no uncertainty as to what agency or party should be responsible for taking action in on-farm management: by definition, action takes place at the farm level.

Land retirement was a source of certain conflict for various organizational activities of the irrigation bureaucracy, even at the initial planning and design stage. The unacceptability of land retirement to the agricultural community was evident in reporting and editorializing about the issue (Eckhouse 1989; Hartshorn 1985). 'To [farmers] land retirement looks a lot like defeat, giving up on a resource they care about. And, more troubling, admitting that irrigating this land was somehow a big mistake' (Hall 1989b).

In the terminology of the sociotechnical matrix, the planning and design of land retirement puts into question the decision making and resource mobilization abilities of the irrigation bureaucracy. Simply as an argument, land retirement pushed the bureaucracy to the limits of its mission, thereby also affecting its conflict resolution and communication activities. And for good reason: environmentalists saw it as no problem at all. As we have already seen, taking irrigated land out of production was for many the preferred solution.

While the land retirement proposal met with certain organizational resistance early on, on-farm management proposals have been developed, and in some cases tested and introduced. Yet, despite construction and O&M activity in the field, efforts to implement these BMPs have been very much bench-scale projects in the planning and design phase (Oster *et al.* 1988; San Joaquin Valley Drainage Program 1987). While bench-scale projects are by definition uncertain, their problematic nature arises primarily because of uncertainties in the organizational activities required to undertake their planning and design. Instead of eliciting stiff bureaucratic opposition, as with the land retirement proposal, the introduction of BMPs caused a great deal of organizational uncertainty, particularly in the areas of decision making and resource mobilization. This is best seen from an investigation of factors that interviewees in irrigation agencies claimed to be constraining the implementation of BMPs.

The causal dependencies between problems related to BMPs were defined in incompatible and conflicting ways by the irrigation experts, as illustrated in the transfer loop in Figure 5.3. The circularity in argumentation illustrates uncertainties in resource mobilization and decision making about BMPs, especially given the strong propensity of many irrigators to want to delay taking any decisive action on the drainage issue – a characteristic that even farmers themselves admitted (Tarr 1988). The introduction of BMPs reinforced the potential for greater uncertainties as well, because none of the proposed on-farm management practices promised a long-term solution to the drainage problem (Imhoff 1989).

Total dissolved solids treatment

Salt separation from drainwater is the best known and most extensively developed of the proposed drainage treatment technologies. Both reverse osmosis and evaporation ponds have been developed through planning and design, construction and O&M, if not in California then elsewhere in the world (Arad and Glueckstern 1981; Lee *et al.* 1988a; Trompeter and Suemoto 1984). Practical experience in decommissioning evaporation ponds was gained during the shut-down of the Kesterson reservoir (Horne 1988; *San Joaquin Valley Drainage Program Status Report* 1987, 1988a, 1989). The following discussion concentrates on evaporation ponds.

Responsibility for construction and maintenance of evaporation ponds has been divided, and became even more so after the Kesterson incident. Management of Kesterson was the joint responsibility of USBR and US Fish and Wildlife Service (Tanji *et al.* 1986), whereas the rest of the valley has both regionally-operated evaporation ponds (such as those in Tulare Lake Drainage District and Lost Hills Water District) and more locally-operated ones at the farm level (Westcot *et al.* 1988). Thus, the party responsible for construction, operation and/or maintenance is fairly clear depending on the geographic location in the valley. What has been missing is an unambiguous and coherent policy for dealing with toxic evaporation ponds in the valley as a whole. The Kesterson incident and subsequent toxicity problems in other evaporation ponds in the valley have shown that reliable management of toxic substances has no tolerance for localized O&M trials and errors. Without a coherent policy, there is little reason to believe that the uncoordinated, geographically fragmented efforts will have any positive, let alone lasting, system-wide impact.

Not surprisingly, decision making on evaporation pond planning and design, construction, O&M and decommissioning has been at a virtual standstill in the valley. Resource mobilization for the decommissioning of evaporation ponds has also been very problematic. This was evident in the difficulties associated with the funding and procedures for the Kesterson clean-up: 'The seemingly settled dispute over how to clean up Kesterson was suddenly reopened last week, when powerful leaders of a House [of Representatives of the US Congress] subcommittee ordered [Secretary of the Interior] Hodel to stop work on the cleanup plan because "there appear to be more effective and less costly options"' (Liebert 1988).

The concern that other evaporation ponds may turn into mini-Kestersons was raised publicly in the press during the debate (Diringer 1988a, 1989; Harris and Morris 1985), and there has been increasing scientific evidence of toxicity problems emerging in the southern valley's evaporation ponds that are similar to those found at Kesterson reservoir (Fujii 1988; Schroeder *et al.* 1988; Westcot *et al.* 1988). As a result, unclear communication about the drainage problem has become a form of conflict management for the agricultural community. From the perspective of officials in the California Department of Water Resources (DWR) and SJVDP, pond operators have by and large been cooperative in providing valuable information to the irrigation bureaucracy on potential adverse impacts of the ponds (personal written communication with the staffs of California Department of Water Resources and San Joaquin Valley Drainage Program on the June draft of Hukkinen *et al.* 1988). Yet a dramatic unwillingness to discuss or even admit evaporation pond toxicity problems was found in the interviews, particularly in the southern San Joaquin Valley. 'I am not aware of any toxics problems in the [Kern County] area', was the response of one water contractor interviewed. By avoiding a full discussion of these problems, such unforthcoming respondents find themselves in the position of calling the mini-Kestersons reported in the press 'a surprise' and a cause for concern and uncertainty over what to do 'now'. These experts would rather cope with the uncertainty caused by not discussing obvious problems, it seems, than run the risk of having the problems publicly discussed in a way that would have made them even more contentious.

Selenium treatment

Since cost differences between the various selenium treatment technologies under investigation in the 1980s were relatively small or altogether unknown, and since even the most advanced of the proposed technologies were only in the bench-scale planning and design stage, the proposed technical solutions will be treated here in the aggregate (Lee *et al.* 1988a).

Agency responsibility for the initial development of innovative selenium treatment technologies was spread throughout all levels of the irrigation bureaucracy. While USBR and DWR were intimately involved in the actual planning and design of selenium treatment technologies, in practice much of the funding of drainage-related research came from SJVDP. In addition, Westlands Water District and Panoche Drainage District were two of the many regional and local agencies responsible for the initial development of such new technologies (*West Valley Journal* 1988c; *Westlands Water District Drainage Update* 1986).

Since planning and design work was scattered throughout the irrigation bureaucracy while on-site construction had not taken place, expectations about which agency or party would eventually be responsible for the O&M of drainage treatment technologies varied considerably. Some thought that selenium treatment would be best handled by a large regional entity (San Joaquin Valley Drainage Program 1987). In its justification for placing selenium treatment at the local district level, the State Water Resources Control Board (SWRCB) report argued that 'individual farmer operation of selenium removal processes does not seem feasible' (Jenkins 1986: 40). Yet some of the interviewees expressed their view that the 'reasons for [our] interest in the Harza [selenium removal] process are that it is simple, it can be operated on-farm.'

But decision making about the planning and design of selenium treatment technologies was not as stalled as it could have been, given the diffuse agency responsibilities. A major environmental group, the Environmental Defense Fund, was actively involved in the initial development of a treatment technology in cooperation with Westlands Water District (WWD) (*EDF Letter* 1987). However, one should not make too much out of this cooperation. The politics of selenium treatment required that there be movement toward dealing with selenium-tainted drainage. It was easier to achieve that cooperation in the planning and design phase than later: implementation and operation were altogether a different matter.

The proposed construction of a 1 million gallon per day (MGD) selenium treatment pilot plant in WWD illustrates the number of potential uncertainties in the decision making and resource mobilization concerning selenium treatment. The particular issue stems from the estimated cost of the proposed plant and the selenium concentration levels achievable by the treatment process. The WWD's position was that the consultant, EPOC Agricultural Corporation, had promised to reach selenium levels below 10 parts per billion (ppb) through biological treatment only, which was cheaper in this case (*Westlands Water District Drainage Update* 1988a). The consultant organization, on the other hand, maintained that it promised to achieve selenium levels below 10 ppb with certainty only if an additional ion exchange unit followed the biological process, thereby increasing total cost (EPOC AG 1987). The attainment of the 10 ppb level by the 1 MGD biological plant alone is possible, EPOC said, but not certain. The development of the biological treatment process as a cooperative effort by the WWD and EPOC eventually 'hit a snag' (*West Valley Journal* 1988a), and WWD had turned its interest and resources to yet another novel treatment technology for selenium, namely an evaporative cogeneration process (*West Valley Journal* 1988b; *Westlands Water District Drainage Update* 1988b).

Yet the selenium treatment process in question was among the most promising ones – SWRCB, for example, used it as the basis for cost estimates of drainage management scenarios in northern San Joaquin Valley (California State Water Resources Control Board 1987). In a more amenable public and bureaucratic environment dealing with more straightforward technical problems, these issues would probably not have been issues at all. Disagreements could have been resolved simply by going back to the conditions set out in the contracts between WWD and EPOC. In practice, however, not only did the inherent uncertainties of the physical phenomena contribute to the complexity of the issue, but it was to the advantage of the irrigation and drainage interests involved to present the issue as a complex one in terms of decision making.

If the public and the bureaucracy had adopted the approach of building biological selenium treatment plants as the preferred solution to the drainage problem, then by implication people would in some sense have accepted the necessity of living with toxics and toxic hotspots. The process in question is one of toxics separation and therefore produces toxics which in turn have to be disposed of in some fashion. The question of public and bureaucratic acceptance of toxics is at the very heart of the post-Kesterson drainage controversy. Californian legislators and voters offered any number of bills seeking to ban toxic hotspots from the waters of the state, such as the Katz Toxic Pits Cleanup Act and the Safe Drinking Water and Toxics Initiative (Proposition 65) during the 1980s. Since people do not want to live with toxics, constructing technologies that make their presence known and their disposal an absolute necessity are likely to cause conflict (Mazmanian and Morell 1988). Thus, while the irrigation bureaucracy has a positive incentive to try to find a solution to the selenium problem as a way of allaying public fears, it has no incentive to adopt a solution that makes selenium disposal a reality.

Disposal

Disposal is the crucial drainage management step. On-farm management options reduce the rate of drainage production, but do nothing to solve the problem of water and salt balance in the soil. In a sense, on-farm management only postpones the time when investments for disposal must be made. Treatment technologies separate the unwanted fractions from the drainage flow, but always produce fractions that must be disposed of.

Technical solutions for drainage disposal are several. The only disposal 'technology' requiring O&M in the valley has been discharge of untreated drainwater into the San Joaquin River. Other technologies are in the planning and design phase, at best, and include conversion of Kesterson reservoir into a landfill, deep-well injection, discharge into the Bay–Delta via a drain, ocean discharge and selenium volatilization in the air (Lee *et al.*

1988a, 1988b; San Joaquin Valley Interagency Drainage Program 1979). All except deep-well injection will be discussed below.

Before the Kesterson incident, the agencies with authority and obligation in drainage disposal were commonly perceived to be USBR and DWR, if not in law then in practice (San Joaquin Valley Interagency Drainage Program 1979). After the Kesterson incident, the expectation changed, with many of the interviewed decision makers perceiving that the responsibility for disposal had shifted to local districts (see also Hall 1989a).

A specific example of uncertainty in agency responsibility for disposal is the state's regulation of drainage discharges to the San Joaquin River. After extensive studies, SWRCB directed the Central Valley Regional Water Quality Control Board to modify its Basin Plan and set standards for some drainage constituents discharged into the river (California State Water Resources Control Board 1987). However, both of the state regulators interviewed (one with SWRCB and the other with the Regional Board) strongly disagreed with this mandate. As one of them put it: 'The drainage problem is not a regulatory problem, but a resource management problem.'

Furthermore, the San Joaquin River regulation was delayed. The SWRCB Technical Committee started working on the proposed regulations for the San Joaquin River in March 1985. According to initial plans, the Regional Board was to have considered the regulatory amendments to the Basin Plan by April 1987 (California State Water Resources Control Board 1986); in reality, the Regional Board adopted those amendments in December 1988, and SWRCB did not approve the amendments until September 1989. Moreover, the amendments were approved only on an interim basis, subject to modification at a later date. In other words, it is the regulators themselves who disagree in decision making over this area. While some disagreement may well have been the normal push and pull of bureaucratic in-fighting, its presence did nothing to make decision making any more assured or certain.

One reason for such disagreements among regulators themselves stems from the trade-off between making a decision and leaving decision making uncertain in the San Joaquin Valley drainage debate. Trying to bring closure to the drainage issue would further mobilize sentiment in the state against irrigation, particularly since the impact of current disposal technologies is either unknown or not agreed upon. The moment the potential for such conflict arises (in this case through regulatory action), mechanisms simultaneously emerge to prevent this polarization (in this case as an unwillingness to regulate, defended by claims that the issue is more complex and uncertain than first meets the eye).

The clean-up efforts of Kesterson reservoir represent both the decommissioning of an evaporation pond and the planning, design and construction of a disposal site, in this case a landfill. Uncertainty in decision making was abundantly manifest at every stage in these activities. The decision over which decommissioning technology to use was confused and thus delayed by the continuous emergence of suggested new technologies, each subsequent technology supposedly a more sound solution than the preceding one. At first, the ponds were to become a monitored landfill; then they were to be flooded with uncontaminated water; later USBR had the ponds filled with uncontaminated soil (Horne 1988; *San Joaquin Valley Drainage Program Status Report* 1988a, 1988b). To make matters worse, SWRCB and USBR continually disagreed over responsibility and resource mobilization in the clean-up (Diringer 1988b; Liebert 1988).

SJVDP's handling of the master drain and ocean pipeline disposal options is a case in point, not just of the decision making dilemma but also of the management of conflict through ambiguous communication. In one of its first publications, the Program proclaimed its aims to 'identify measures to help solve immediate drainage and related problems and to develop comprehensive plans for long-term management of those problems' (San Joaquin Valley Drainage Program 1987: 2). Yet it delimited these goals a few pages later by stating that, based on directives from the Program management and advisory committees, 'no studies of out-of-valley disposal of drainage water are planned to be conducted by the Program' (San Joaquin Valley Drainage Program 1987: 27). A later report by SJVDP stated that 'because of its sheer magnitude the salt problem cannot be solved, over the long term, solely in the valley' (Imhoff 1989: iii).

As a result of political pressures, SJVDP staff was not at all determined to fulfil its basic task of developing a comprehensive, long-term drainage management plan for the San Joaquin Valley. Despite its fundamentally short-term goal-setting, SJVDP spent a good 50 million US dollars on drainage-related research (Hall 1989a). On the one hand, the Program's avoidance of controversial research on out-of-valley solutions kept alive the uncertainty over the feasibility of such solutions. This avoidance was also noted in a letter from the National Research Council's Advisory Committee to SJVDP, which criticized the Program's management for restricting its scientists to those options considered 'politically feasible' (National Research Council 1989; *San Francisco Chronicle* 1989). On the other hand, the Program's research on in-valley solutions increased uncertainties by producing findings in areas unfamiliar to irrigation bureaucrats and irrigators. But no individual or group could polarize the issue by accusing SJVDP of not being active in trying to solve the drainage problem. Again, uncertainty plays a functional role for an agency intent on avoiding any further polarization of the drainage issue, which would jeopardize the irrigation bureaucracy's long-term survival.

Lastly, the option of letting selenium volatilize into the air is similarly problematic, but in a slightly different way. Here the uncertainty revolves around how this option was communicated. SJVDP, others in the irrigation bureaucracy and their consultants (such as university researchers) articulated volatilization not as a disposal option *per se*, but as 'a natural biological process common in all ecosystems', which 'recycles [selenium] into the atmosphere' (San Joaquin Valley Drainage Program 1987: 25). Yet why were all other disposal options not also cast in this language of natural as distinct from artificial processes? Why is recycling selenium by air any more 'natural' than recycling it by other disposal methods, such as discharge into the San Joaquin River? Obviously, there may have been good political and bureaucratic reasons for this bit of obfuscation. But the decision to make rather arbitrary and ambiguous discriminations between disposal processes only added to, rather than reduced, the uncertainty and complexity of the drainage debate.

REDEFINING EXPERT SYSTEMS

The California Central Valley drainage problem is an inherently systemic one. Its source is neither a conspiracy by a few politically astute individuals, as has often been implied by researchers of Californian water policy (Reisner 1987); nor is the problem restricted to any one of the agencies involved in drainage management. Instead, the fundamental drainage problem stems from the overall organizational dynamics, which determine the way key decision makers in the state's water agencies perceive, define and approach every drainage-related issue.

The systemic nature of the valley drainage problem explains why it has been, is, and will be impossible to locate the problem or its remedies through conventional technical and managerial planning methods, which typically take the existing organizational structure as given, and solve problems by adjusting the technical and economic features of proposed solutions to meet resource constraints. It also explains why the interviewees failed to evaluate the feasibility of potential technical solutions: not only were the management alternatives too uncertain, but more importantly, the experts were incapable of even considering the feasibility of a preferred option. Instead, they were overwhelmed by speculation about what effects the option, if implemented, would have on organizational survival and authority. Finally, the systemic nature of the problem dictates that its remedies deal with the institutional sources of the decision makers' cognitive dilemma.

The Bayesian expert system used in this study provides a useful reformulation of the long-standing drainage problem in the San Joaquin Valley. The analysis indicates a way toward a valley-wide and long-term remedy. By administratively decoupling drainage from irrigation allocation and distribution, the drainage problem is redefined in organizational and political terms, which have long been accepted as crucial parameters of the issue but have failed to play a substantial role in the proposed remedies. The call for organizational restructuring of the valley's irrigation and drainage management does not lessen the importance of techno-economic and ecological parameters, but rather suggests that techno-economic intervention must be facilitated by simultaneous administrative reform.

Methodologically the expert system constructed is unique in at least three aspects. First, it is used as a problem definer rather than a problem solver. Traditional expert systems combine decision rules by experts in a given area into a deduction system capable of solving a range of problems. In the San Joaquin Valley drainage case, however, the expert system is an aggregation of individual experts' causal explanations of the problem. Bayesian analysis can effectively expose the essential elements and relations in the causal problem networks. As Hart *et al.* (1984) long ago noted, the overwhelming complexity and uncertainty of many environmental issues has made the inability to construct well-formed problems the chief obstacle to policy making.

Second, when the problem-focused expert system is designed, it is not necessary to harmonize the inconsistent conceptual frameworks used by individual experts. Instead, these inconsistencies become the starting point for understanding the problem in a way that makes it more amenable to remedial action. In this way any recommendation for remedy remains true to the original expert statements rather than building upon selectively adopted or reformulated expert statements.

Finally, formulation of the Bayesian expert system is not restricted to empirical explanations, but allows normative argumentation as well. Drainage facts do not alone determine how experts operate. Their actions are also guided by organizational and political motivations, which years of manoeuvring within the water institutions have moulded. As a result, an expert cannot view the problem solely as an objective, disinterested analyst, but is inevitably persuaded to define the problem and its potential solutions through the lens of relevant political considerations. An academic review panel found evidence of such behaviour in its evaluation of the San Joaquin Valley Drainage Program activities, particularly the Program's refusal to consider out-of-valley disposal options, such as the master drain (National Research Council 1989). In the Bayesian analysis of the drainage problem, explication of the experts' norms and political beliefs provided crucial insights on the problem. Such has been the experience from other politically charged policy issues as well (Nelkin 1984).

The applicability of network analysis is not restricted to agricultural drainage problems in the American West. Environmental problems increasingly appear as a seemingly intractable mixture of factual uncertainty and political controversy. As the next two chapters will show, the method can be fruitfully applied in very diverse environmental management issues, not just in the US but around the world.

6

CORPORATISM AS AN IMPEDIMENT TO SUSTAINABLE WASTE MANAGEMENT IN FINLAND

The third case study, focusing on Finnish waste management, expands the institutional argument presented in Chapter 2. Where the first two cases dealt with irrigation-related environmental problems in arid areas, we now move to solid and hazardous waste management issues in boreal regions. The Finnish case elaborates the feedback between expert models and formal institutions at three levels of institutional rule: operational, collective choice and constitutive. Formal institutions at all three levels lead decision makers in Finnish waste management to allow short-term economics to guide their decisions, despite their preference for environmentally sustainable waste management. The pressure from formal institutions increases the dissonance between a decision maker's preferences and the institutional constraints. The individual strives to reduce the dissonance by following – and thus reinforcing – the formal institutional setting. I will pay particular attention to what I call environmental corporatism, which turns out to be the fundamental, constitutive rule that guides environmental policy makers in Finland to consistently prioritize short-term economic imperatives over longer-term environmental ones.

Several authors have argued that the corporatist state has the potential to excel in long-term economic performance and industrial adjustment. Corporatism can be crudely characterized as the institutionalized integration of conflicting constituencies in decision making. Non-exclusive corporatism, which institutionalizes bargaining mechanisms in a comprehensive fashion among all conflicting interest groups, is seen as a particularly promising guarantor of strategic policy (Landesmann and Vartiainen 1992; Pekkarinen *et al.* 1992). But the case study on Finnish waste management indicates that policy makers may in the long run be bitterly disappointed to see corporatism destroy the ecological foundation of economic performance.

The aim of the study on which this chapter is based was to improve the capacity of decision makers in Finland, a non-exclusive corporatist society *par excellence*, to design and implement long-term waste management policies.

The central finding is that while inclusive corporatist negotiations may be far-sighted from the point of view of national economic and industrial performance, they have the systemic characteristic of excluding issues of long-term ecological sustainability: corporatist institutions nurture environmental policy makers' short-sighted mental models, which in turn are the rationale for the short-sighted corporatist institutions. Interviews indicate that despite the high awareness among Finnish policy makers of the need to take into account the long-term future, the institutional framework of corporatism is a disincentive for them to work out and implement environmentally sustainable policies. Finnish corporatism prevents ecological sustainability from even being seriously considered in corporatist negotiations, because decision makers themselves conceptualize environmental issues in unproblematic terms. The main long-term environmental policy problem in Finland is therefore not environmental conflict, but its absence. To improve the situation, corporatist institutions should be dismantled to allow the policy makers' latent ecological awareness to shape long-term policies.

WASTE MANAGEMENT AND ITS ADMINISTRATION IN FINLAND

Public agencies and private firms at national, provincial and local levels make strategic decisions on Finland's waste management. The Ministry of the Environment is the chief policy maker and regulator of municipal and hazardous waste management. Provincial governments permit and monitor both municipal and industrial waste management at the regional level. Municipal governments plan and regulate municipal waste management at the local level, but also arrange the transport, disposal and recycling of municipal waste. As a rule, they have subcontracted waste transportation to private waste management firms. Industrial and mining processes are the source of approximately half of the 65 to 70 million metric tons of waste generated annually in the nation; the other half is agricultural (34 per cent), construction (10 per cent) and municipal waste (6 per cent). Approximately 45 per cent of the total waste stream is recycled (Koivukoski 1992; Palokangas *et al.* 1993; Vahvelainen and Isaksson 1992).

As in any modern industrialized country, hazardous wastes have received particular attention in Finland. The country's hazardous waste management problems have much to do with its large land area and small population, leading to numerous small landfills (a total of 680 in 1990). Many have been closed down, but this has not diminished concerns that the closed landfills may in fact be environmental crises in the making due to storing of hazardous wastes generated during the country's rapid industrialization since the 1950s (Assmuth *et al.* 1990). The situation improved considerably in 1984 when the nation's centralized hazardous waste management plant, Ekokem, began its operations. The central government, municipalities and industry own the plant jointly. When this case study was conducted in the early 1990s, a total of 233,000 metric tons of hazardous waste was generated in Finnish industry, of which approximately 100,000 metric tons was recycled as raw material, 90,000 treated at waste water treatment plants and 43,000 treated or stored otherwise. Of the 43,000 metric tons, 52 per cent ended up at Ekokem (Vahvelainen and Isaksson 1992).

The objectives of Finland's waste management policy are similar to those adopted in other industrialized countries, namely, to reduce the amount of waste generated. What remains is then to be recycled, recovered or reused to the extent feasible. That waste which can be neither reduced nor recycled is to be disposed of in an environmentally acceptable fashion (Office of Technology Assessment 1989; Rudischhauser 1992; RVFs Framtidskommitté 1991; Waste Management Advisory Board 1991; Waste Management Council 1992). As the following illustrates, the administrative structure of Finnish waste management and the institutions that support it largely prevent the realization of these policies in their stated order of priority.

INSIDE THE CIRCLES OF ENVIRONMENTAL CORPORATISM

My methodological approach was to analyse the structure, substance and institutional context of the mental models with which key policy makers in Finnish waste management rationalize their long-term decisions. I considered individual scenarios as equal elements of a collective and often internally inconsistent model of the future. Analysis of the factual, attitudinal and logical inconsistencies of the collective model indicates how formal institutions constrain long-term policy decisions and what institutional changes might relax the constraints.

To map the cognitive models underlying waste management policies, I conducted interviews in May and June of 1992 with twenty-four decision makers and experts of Finnish waste management. The interviewees were classified into five interest groups based on their employment: five were university researchers (researchers), four worked in consultancies (consultants), another four represented private firms or industries (entrepreneurs), six were planners or regulators in the public sector (bureaucrats), and five were politicians or representatives of special interest group organizations (politicians). The loosely structured interviews focused on two issues: first, what would become the most pressing waste management problems in Finland in the next fifty years; and second, what challenges the long-term threats posed for today's decision making. 'Long term' could, of course, just as well have been forty or sixty years – it is the order of magnitude that matters. That said, five decades into the future is close enough for the interviewees to feel they can somehow influence it, while far enough to exceed the average lifetime of the current waste management infrastructure. The interviewees were therefore unable to fall back solely to the more familiar territory of assessing the shorter-term effects of normal investment decisions.

No less than 282 different problem statements were identified in the twenty-four interviews. The problems and the linkages the interviewees stated between them were coded into problem networks, as described in Chapter 3 (see pp. 41–4). Individual problem networks were aggregated at the interest group level to find out the differences and similarities in the mental models of the five groups. As in the Californian case study, the focus of analysis was on loops and terminal paths resulting from loops. Loops are prevalent in the problem networks of researchers, consultants, politicians and bureaucrats. In contrast, entrepreneur networks are linear, with initial problems leading to a few terminal problems via several transfer problems. This difference between cyclic and linear networks plays a central role in the interpretation of the cognitive maps.

What is striking about the Finnish case is that circular argumentation predominantly originates at individual level. Intra-group aggregation of individual networks typically enriches the loops mentioned by individuals. An individual's loop $1 \leftrightarrow 2$, for example, may, when aggregated with the networks of other individuals of the same group, transform into $1 \rightarrow 2 \rightarrow 3 \rightarrow 1$. The emergence of loops at individual level sets the Finnish case in interesting contrast to the Californian case, where the loops primarily arose when the problem networks of individuals belonging to two different interest groups were aggregated. This is peculiar, since organization theory would lead us to expect individuals not to justify their actions with circular arguments (Simon 1964). One explanation may be essential differences in what the experts are discussing. In the Californian case study, the environmental problem was one requiring urgent action, which would have been impossible without a clear distinction between causes and effects. The future of Finnish waste management is altogether another matter, in which the absence of urgency leaves plenty of room for a more casual pondering upon causes and effects. The important thing to note, however, is that in both case studies cognitive dissonance at individual level was observed. In the Californian case study, it was evident in the operating assumptions the experts perceived to be constraining potential solutions to the drainage problems. In the Finnish case study, as will become clear soon, an individual expert's circular argument was logically possible only with cognitively dissonant thinking.

The mapping of the cognitive models that policy makers in Finnish waste management use to understand the future can be summarized in the following findings:

1 Waste management experts and decision makers typically describe future waste management problems as loops. Fourteen of the seventeen loops that emerged in network aggregation by interest group were mentioned by individuals, and half of the twenty-four individuals mentioned loops. Loops are held together by a cognitive goal conflict between profit-maximizing goals, which prioritize short-term economic profit, and sustainability goals, which aim at the preservation of ecosystems over generations. Goals are here understood as normative statements about the state of affairs individuals or organizations would like to achieve (Scott 1987).
2 The goal conflict does not, however, interfere with the interviewees' day-to-day decision making. All of the loops indicate that expert advice and decisions are guided by profit maximizing, short-term operating assumptions. Operating assumptions reflect what individuals or organizations consider achievable in society under existing constraints and conditions (Hukkinen *et al.* 1990).
3 The loops also indicate that the profit-maximizing operating assumptions are institutionally rooted in the administrative, technological, economic and political structures of the Finnish society. This phenomenon will be referred to as environmental corporatism. It has the systemic property of integrating conflicting environmental policy interests to the extent that their open political resolution is impossible. In the administrative sphere environmental corporatism fuses together implementation and regulation in waste management. In the technological sphere environmental corporatism obfuscates the economic and political conflicts of interest that underlie different stages of waste management technology by operationalizing waste management as a harmonious flow of engineering operations. In the economic sphere environmental corporatism ensures that all environmental problems are treated as short-term microeconomic questions. And the political sphere of environmental corporatism is dominated by politicians who interpret the general public to prefer short-term economic growth to long-term ecological sustainability.
4 Finally, the interviewees foresee a gloomy long-term future resulting from their adherence to profit-maximizing operating assumptions. Forty of the fifty-four terminal problems identified in the aggregated networks paint a picture filled with threats to the existence of Finnish waste management organizations, society and ecosystems.

Every one of the seventeen loops and their consequences supports the four findings. Tables 6.1 to 6.5 summarize the contents of the loops and terminal problems and briefly describe how they support the four findings. The following analysis focuses on one loop and its consequences in the researchers' aggregated problem network, another loop and its consequences in the bureaucrats' aggregated problem network, and a linear problem path in the entrepreneurs' aggregated problem network. Case study examples from municipal waste management and hazardous waste management support the analysis and illustrate in detail the mechanism of interaction between the perceptions of the key formulators of Finnish waste management policy and the institutional framework within which they operate.

Researchers' loop 1 and its consequences

According to loop 1, which was described by interviewee no. 3, natural resources are becoming ever scarcer in today's economy. This makes the need for sustainable waste management policy all the more urgent. Such a policy, however, would reduce material standard of living and therefore run against efforts to revitalize the stagnant economy. In interviewee 3's experience, revitalization policy wins, which closes the loop when natural resource consumption increases further.

The circular argument in loop 1 is possible because of the interviewee's simultaneous adherence to profit-maximizing and sustainability goals

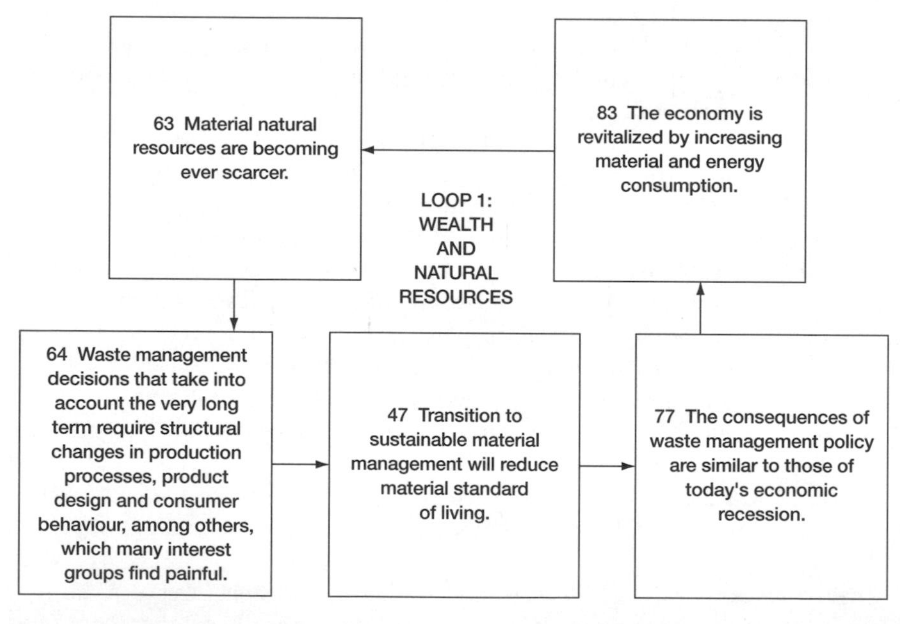

Figure 6.1 Loop 1 by researchers in the Finnish waste management case study. Economic recession is perceived to be increasing pressures to stimulate material consumption, which in turn is seen to be increasing pressures to reduce material consumption to achieve sustainable waste management.

(finding 1). It is only by adopting conflicting goals that the interviewee can regard as problems both the possibility that transition to sustainable waste management means reducing material good (problem statement 47) and that economic revival means increasing material good (problem statement 83). However, decisions on waste management are being made with reference to short-term profit-maximizing operating assumptions that emphasize material economic growth (finding 2). Despite clear affinity to sustainability goals (as expressed in problem statement 64, for example), interviewee 3's operating assumption is that the economy will revive (problem statement 83) and that natural resources will become ever scarcer (problem statement 63).

Loop 1 also offers some indication of the institutional factors behind the belief that profit maximizing will dominate decision making about waste management (finding 3). First, interviewee 3 believes that political decision makers operate on the assumption that the general public ultimately wants more material well-being (the chain of reasoning from problem statement 64 to 83). Second, the interviewee believes that the technological production systems from production processes to product design are structured to support an economic system that accumulates material wealth (problem statements 64 and 83). Finally, the dominant economic policy is in the interviewee's opinion inclined to correct its dysfunctions by increasing material production (problem statements 77 and 83).

The interplay between perceptions expressed in researchers' loop 1 and the institutions of environmental corporatism becomes clear by investigating the early-1990s dispute over the construction of a municipal waste incinerator to ease the Helsinki region's waste management problems. Opponents of the incinerator argued that it would nullify efforts to reduce and recycle waste. Proponents of the plant argued it was a way to recover the energy content of waste (Pohjanpalo 1991). In the end, the board of the semi-governmental waste management authority decided to postpone final decision until the next decade (Nousiainen 1992).

Now consider the political debate over waste incineration in light of the above-described loop by researchers and loop 3 by politicians (Tables 6.1 and 6.2). The central message of politicians' loop 3 is that the inclusion of strong interest groups in political and administrative decisions on waste management is the main reason for short-sighted policy, but also the prerequisite for reaching any decision at all. The Helsinki region's semi-governmental waste management authority contains a full assortment of interests, including an engineering staff responsible for the technical implementation of waste management and a politically elected governing board responsible for waste management policy. The important characteristic is the institutionalized mixing of implementation and policy making. During the debate over waste incineration the policy makers on the board were under constant lobbying from the technical staff, who supported waste

Table 6.1 Summary of the cognitive mapping of researchers in the Finnish waste management case study

Network element	*Evidence for proposition 1 (conflict between short- and long-term goals)*	*Evidence for proposition 2 (dominance of short-term profit assumptions)*	*Evidence for proposition 3 (environmental corporatism)*	*Evidence for proposition 4 (threat scenarios)*
Loop 1: Economic recession increases pressures to stimulate material consumption, which in turn increases pressures to reduce material consumption to achieve sustainable waste management.	In the long run, revitalization of material consumption is a problem, but in the short run its absence is a problem.	The economy is revitalized by increasing material and energy consumption.	The techno-economic system is programmed to correct itself by increasing material consumption.	
Loop 2: Increased material well-being generates more waste, which in turn diminishes material well-being and leads to further demands to increase material well-being.	From the short-term viewpoint the reduction of material standard of living is a problem, from the long-term viewpoint it is not.	Political pressures lead to policies that increase material well-being.	The political and economic decision making system corrects social problems by increasing material well-being.	
Loop 3: Belief in market forces as the remedy to environmental ills lures businesses to fraudulent practices, which only reinforce the fierce market competition.	From the long-term ecological point of view excessive belief in the market is a problem, but from the short-term economic point of view environmental subsidies are a problem because they distort competition.	The market, commercial effectiveness and privatization are trusted as solutions to waste management problems.	Decisions in both public administration and in private corporations are based on micro-economic models.	
Terminal problems resulting from researchers' loops 1 to 3.				Widespread absence of strategic thinking in the society; inability among corporations to bring about ecological restructuring; permanent negative changes in ecosystems; environmental wars.

Source: Hukkinen 1995a: 71.

Table 6.2 Summary of the cognitive mapping of politicians in the Finnish waste management case study

Network element	*Evidence for proposition 1 (conflict between short- and long-term goals)*	*Evidence for proposition 2 (dominance of short-term profit assumptions)*	*Evidence for proposition 3 (environmental corporatism)*	*Evidence for proposition 4 (threat scenarios)*
Loop 1: Investment in centralized waste treatment plants closes out step-by-step options and emphasizes centralized waste disposal over waste reduction.	In the long run it is questionable to commit resources to centralized plants, but in the short run waste needs to be treated somehow.	Resources are invested in centralized waste management systems that close out other options.	Decision making over waste management policy is blind to the sociopolitical conflicts between waste disposal, recycling and reduction.	
Loop 2: The lack of international agreements on waste recycling slows down the development of recycling technology, which discourages international agreements.	In the long run countries should develop ecologically sound production technologies, but in the short run an individual country cannot change its production systems.	Ecologically sound production technologies and international agreements promoting their adoption do not exist.	Finnish politicians interpret the realism of international environmental politics so that Finland alone cannot begin ecological restructuring of industrial production.	
Loop 3: The inclusion of strong interest groups in waste management decisions prevents long-term policy, but their exclusion would prevent any decisions at all.	Environmental officials realize that sustainable waste management is necessary in the long run but avoid such policies for the sake of short-term political expediency.	Politicians and bureaucrats do not create long-term waste management policy but rather focus on day-to-day disputes over the formulation of statutory text.	To secure a feasible waste management policy the political and administrative decision making system incorporates the short-term views of strong interests.	
Loop 4: The rigid technological, organizational, political and attitudinal structures of waste management support and are supported by the short-term decision making mechanism.	The inability of current decision making system to consider long-term issues in waste management is acknowledged, but dismantling of the structures that support the system has yet to start.	The technological, organizational, political and attitudinal structures of waste management remain rigid.	The rigid technological, organizational, political and attitudinal structures uphold a decision making system incapable of taking long-term decisions.	
Terminal problems resulting from politicians' loops 1 to 4.				Ecological restructuring of industrial production loses under sociopolitical opposition and lack of profitability.

Source: Hukkinen 1995a: 72.

incineration, and critical environmental groups, who were opposed to incineration (Pohjanpalo 1991). In the end, the policy makers did reach a decision, but only the short-sighted one of avoiding the key issue for the time being.

The dispute over waste incineration also sheds light on how the technological and economic spheres of environmental corporatism intersect with the perceptions of the decision makers. The waste management authority has comprehensive responsibilities that reflect the nation's overall waste management objectives to reduce, recycle, recover and dispose of waste. While these tasks are sensible from the point of view of material flow engineering, their rationality fails under short-term economic and political pressures. The recycling of waste, for example, necessarily reduces the raw material of a waste-to-energy incineration plant. Without guarantees of a continuous supply of waste material, the recycler is reluctant to undertake significant long-term investments. Conversely, investment in a waste recycling facility will undermine waste-to-energy policy. A similar conflict separates waste reduction from energy recovery. The short-term imperative of securing an adequate waste stream to an incineration plant can systematically undermine the long-term issue of creating social institutions that would reduce that waste stream.

While it is true that major investments in technology always direct future choices, the influential power of technology is even stronger under corporatist institutions. On the face of it, the administrative integration of technologically coupled operations under the auspices of a single waste management authority makes sense. But it also corners the decision makers into a cognitive dilemma, when the pursuit of the technologically defined objectives of waste management polarizes the political and economic conflicts of interest that underlie the different stages of waste management technology (Figure 6.1). The same dilemma is also evident in researchers' loop 2 (Table 6.1), politicians' loop 1 (Table 6.2), and bureaucrats' loop 4 (Table 6.3).

Bureaucrats' loop 3 and its consequences

The bureaucrats' loop 3 (Figure 6.2) describes how Finland's hazardous waste disposal facility Ekokem was originally designed to have excess disposal capacity (problem statement 19). This exacerbated the conflict between waste reduction and disposal (problem statement 4). To dampen such criticism, environmental regulators began directing all of the nation's hazardous waste to Ekokem, including waste which could just as well, both environmentally and technically, have been treated elsewhere (problem statement 32). The regulatory policy bore fruit to the extent that Ekokem has repeatedly reached the limits of its disposal capacity (problem statement 33), thus forcing the plant to expand its operation (problem statement 19).

Table 6.3 Summary of the cognitive mapping of bureaucrats in the Finnish waste management case study

Network element	*Evidence for proposition 1 (conflict between short- and long-term goals)*	*Evidence for proposition 2 (dominance of short-term profit assumptions)*	*Evidence for proposition 3 (environmental corporatism)*	*Evidence for proposition 4 (threat scenarios)*
Loop 1: Small and short-sighted research projects lead to technologies, decision systems and attitudes that reinforce short-sighted knowledge production.	Although the need for sustainable waste management policy is acknowledged, knowledge production remains short-sighted.	There is a general reluctance among various social interest groups to sketch out the long-term issues in waste management policy.	Short-term decisions and research reinforce each other so that sustainable waste management does not emerge as a social problem.	
Loop 2: Short-sighted politicians do not develop visionary scenarios, which in turn reinforces their own short-sightedness.	Although politicians acknowledge the need for sustainable waste management, their attention spans only one electoral term.	Sustainable waste management does not materialize as political decisions.	Short-term decisions and research reinforce each other so that sustainable waste management does not emerge as a social problem.	
Loop 3: Ekokem plans excess treatment capacity only to find the capacity inadequate when environmental regulators order more hazardous wastes to the plant.	From the ecological point of view, wastes to Ekokem could be allowed to disappear, but from the economic point of view Ekokem needs hazardous wastes.	Ekokem's existence is secured through administrative orders.	Ekokem survives through the alliance between waste management implementors and regulators.	
Loop 4: Ekokem raises waste fees when the incoming waste stream diminishes, but this diminishes the waste stream and increases the pressure to raise fees even more.	Increased waste fees reduce waste, but are economically unfounded.	Waste streams are secured through the construction of large waste treatment plants.	Large-scale waste management systems sustain waste streams.	
Loop 5: See loop 2 in Table 6.1.	See loop 2 in Table 6.1.	See loop 2 in Table 6.1.	See loop 2 in Table 6.1.	
Loop 6: See loop 1 in Table 6.4.	See loop 1 in Table 6.4.	See loop 1 in Table 6.4.	See loop 1 in Table 6.4.	
Terminal problems resulting from bureaucrats' loops 1 to 6.				Conflicts between corporations and government; permanent disruption of ecosystems; environmental wars.

Source: Hukkinen 1995a: 73.

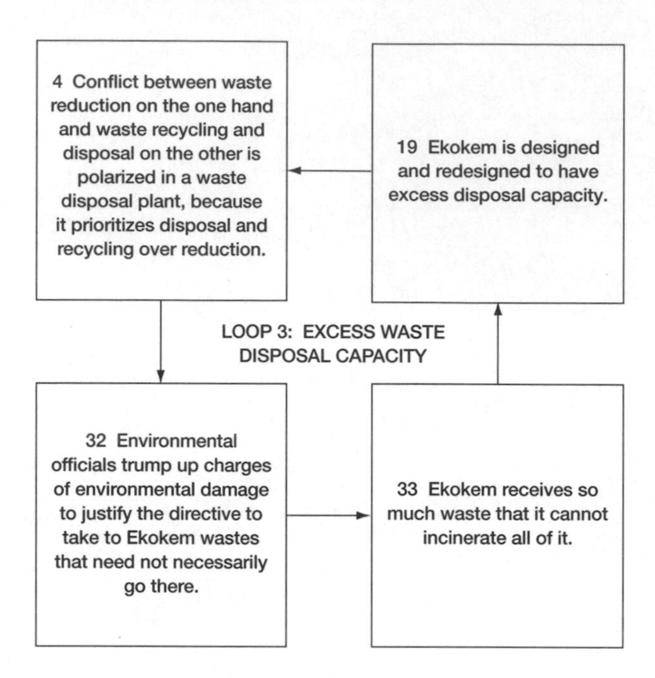

Figure 6.2 Loop 3 by bureaucrats in the Finnish waste management case study. Ekokem is perceived to be in a cycle of planning excess treatment capacity only to find the capacity inadequate when environmental regulators order more hazardous wastes to the plant.

Loop 3, which was stated by interviewee no. 1, supports finding 1. The circularity of the argument arises from the fact that decision makers attempt to deal with both profit-maximizing and sustainability goals. The problem from the profit-maximizing point of view is that the profitability of Ekokem's hazardous waste disposal could suffer if low-waste technologies were to reduce significantly the incoming waste stream. From the sustainability perspective, the threat is that Ekokem's environmentally reliable hazardous waste disposal operations might lose business to competing but environmentally inferior disposal options (problem statements 4 and 32).

Ekokem's officials are in a cognitive double bind. If their only goals were short-term monetary ones, Ekokem's diminishing profits would not be a problem, because hazardous waste could be left to the market to handle. If sustainability goals were the only relevant ones for the regulators, they might view the disappearance of Ekokem as a signal of the initial stages of industrial ecology. However, the simultaneous requirement of short-term economic profit and long-term ecological sustainability puts the officials in a dilemma. Without significant investment in research and development, nothing guarantees that market-based hazardous waste management will also be environmentally sound, or that firms operating in an industrial ecology will also be profitable.

Interviewee 1 believes that profit-maximizing goals will win in the mechanism of loop 3 (finding 2). Ekokem is described as an automaton which was designed for and will continue to work at excess capacity. The interviewee's arguments illustrate how the administrative sphere of environmental corporatism ensures the operation of the automaton (finding 3). Implementors and regulators are perceived to administer Finnish hazardous waste management in intimate cooperation. According to interviewee 1, it is only by rule of environmental regulators that hazardous wastes end up at Ekokem (problem statement 32).

A closer look at the way Ekokem is operated and administered provides a better understanding of the mutually reinforcing relationship between corporatist institutions and policy makers' beliefs. Ekokem has indeed the administrative configuration to secure an adequate supply of hazardous waste. The Ministry of the Environment is the chief regulator of the nation's hazardous waste management, but also owns a third of Ekokem and has a representative on its governing board (Koivukoski 1992). As the Chief Executive Officer of Ekokem noted in a speech following the first full year of operation in 1985, 'The success of [Ekokem] is highly dependent on advisory and regulatory work by local waste management officials and on compliance with waste management law' (Koivukoski 1992: 62). The confusion between implementation and regulation may conveniently secure the short-term operation of the plant, but it is a serious environmental policy problem in the long run. The implementor's primary objective is to manage waste in the best technical and economic manner possible. The regulator's primary interest is to ensure ecologically acceptable waste management in the long run. When the conflicting interests are administratively melded, bureaucrats find themselves in cognitive bewilderment, when fulfilling one administrative duty violates the other (Figure 6.2). Descriptions of related administrative dilemmas can be found in politicians' loop 3 (Table 6.2) and consultants' loop 4 (Table 6.4).

It should be emphasized that the individual regulator is not to be blamed. The short-term environmental policy emanates from the institutional arrangement and does not reflect an individual regulator's lack of commitment to sustainable development. He or she has simply made informed choices under the boundary conditions of the policy machinery. Policy makers in hazardous waste management operate under considerable cognitive dissonance over what they think ought to be the environmentally sustainable policy and what they think is the feasible environmental policy.

The case of Ekokem also illustrates how successful short-term regulation can veil a fundamental incapacity to set long-term regulatory policy. Ekokem has a good record of environmental performance. The Ministry of the Environment is so dependent on budgetary and legislative support from

Table 6.4 Summary of the cognitive mapping of consultants in the Finnish waste management case study

Network element	*Evidence for proposition 1 (conflict between short- and long-term goals)*	*Evidence for proposition 2 (dominance of short-term profit assumptions)*	*Evidence for proposition 3 (environmental corporatism)*	*Evidence for proposition 4 (threat scenarios)*
Loop 1: Every time a new waste management system is created, it inherits the rigid technological, organizational, political and attitudinal structures of its predecessor.	New dynamic institutional structures should be created, but the old rigid structures have their benefits as well.	The new waste management institutions replicate the rigid characteristics of the old ones.	Since all reforms take place in the context of existing rigid institutions, there is no reason to expect the new institutions to be dynamic.	
Loop 2: Poor interaction between researchers and implementors in waste management prevents practical solutions, which alienates the two groups even more.	Long-term waste management solutions require cooperation between environmental researchers and technical implementors, which they are reluctant to do in the short run.	From the point of view of avoiding conflicts in the short run, researchers and implementors benefit from the lack of cooperation.	The political benefits of avoiding research cooperation between environmental and technical sectors are perceived to exceed the benefits of cooperation.	
Loop 3: Economic problems force a waste recycler to increase fees, which reduces the incoming waste stream and pushes waste fees even higher.	In the long run the disappearance of waste recycling capacity is a problem, whereas in the short run the poor profitability of waste recycling is a problem.	Short-term goals predominate and the waste recycler keeps on raising its fees.	Waste recycling is successful only if its inputs and outputs are linked with other economically profitable activities.	
Loop 4: An environmental regulator solves problems in one environmental sector by dumping them onto another sector.	Regulators solve sectoral environmental problems in the short run, but roll over problems to other sectors in the long run.	The regulators' primary operating assumption is to solve narrowly defined short-term problems.	Environmental bureaucracy justifies its existence by recreating problems in one sector after another.	
Terminal problems resulting from consultants' loops 1 to 4.				Cognitive, administrative and political inability to resolve waste problems; international environmental wars.

Source: Hukkinen 1995a: 70–71.

Parliament that it has forced Ekokem to carry out extensive monitoring programmes. Furthermore, the day-to-day regulatory powers over Ekokem rest largely with the provincial authorities who are not directly bound by the ministerial ownership of the plant. The meeting of regulatory standards by Ekokem is not, however, the main issue at hand. It is rather the direction of the nation's overall regulatory policy in the long run. This, unfortunately, is the sole responsibility of a ministry whose long-term vision is seriously obscured by its more immediate concern for the profitability of the waste treatment plant it owns.

Implementation and regulation have been integrated in Finnish waste management planning as well. The Waste Management Plan has been the most important regulatory instrument since the late 1970s. The waste producer must submit a plan for approval by the municipal government (in some cases by the provincial government) whenever the waste is in large quantity, unusual or difficult to handle, or hazardous (Palokangas *et al.* 1993). (The new Waste Law of 1993 [Waste Management Law Committee Report I 1992] changed the permit system somewhat.) The plan is prepared in close cooperation between the firm seeking approval of the plan and the regulator approving it. In fact, planning guidelines by the Ministry of the Environment encourage such cooperation (Ministry of the Interior 1983). The arrangement may in the short run have guaranteed the rapid preparation and approval of a large number of plans throughout the country, but it entails several long-term problems. The administrative mixing of conflicting regulatory and implementing interests in the planning process and the lack of formally stipulated procedures for settling them can easily erode the public's confidence in the authority and independence of the regulator. Furthermore, it draws the attention and resources of municipal and provincial regulators away from long-term strategic issues in waste management to the operational details of an individual enterprise. Finally, the ambition to get as many plans approved as possible through close consultation between regulators and enterprises has created misleading indicators of success in waste management policy. The Ministry of the Environment and the provincial regulators, for example, have measured the success of their policies by the number of approved plans. As a result, continuous and long-term monitoring of the ecological state of the environment, which is the most telling indicator of regulatory success, has suffered from scarce resources (Waste Management Advisory Board 1991).

Entrepreneurs' problem path 1

Figure 6.3 presents a part of the entrepreneurs' problem path 1 that contains common problem statements with the loops described earlier. This enables us to examine the puzzle of how it is that entrepreneurs can perceive as clear-cut linear causalities phenomena that other interest groups

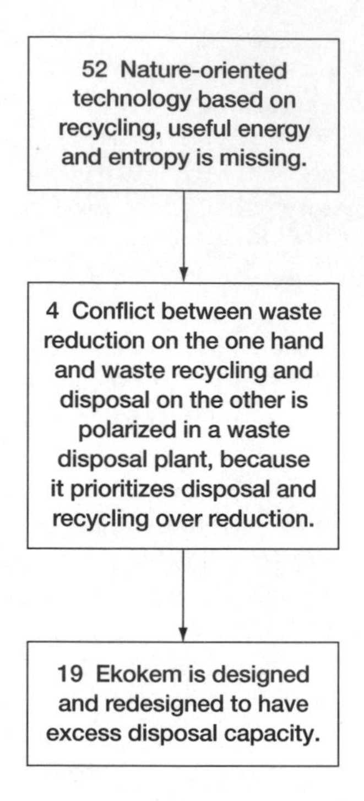

Figure 6.3 Problem chain 1 by entrepreneurs in the Finnish waste management case study. Waste disposal is perceived to be systematically prioritized over waste recycling and reduction in waste management policy.

view as vicious circles. From the point of view of problem solving, a linear problem chain is ideal: solution of the initial problems is likely to solve the terminal ones as well. According to entrepreneurs' problem chain 1, the lack of ecologically sustainable production technologies polarizes the conflict between waste reduction and disposal. As a result, Ekokem, for example, survives only by designing excess capacity (Figure 6.3). The same problems appear in the bureaucrats' loop 3 (Figure 6.2). The difference in argumentation between entrepreneurs and bureaucrats is that entrepreneurs see no need to maintain artificially Ekokem's excess disposal capacity by resorting to, say, decisions by regulators.

Goal conflicts between waste reduction and disposal typically become polarized in governmental and quasi-governmental agencies such as Ekokem or the Helsinki region's waste management authority, which are constantly exposed to the uncertainties of their institutional surroundings. Ekokem's technical directors make decisions under the immediate sphere of influence of regulators from the Ministry of the Environment. Technical staff at the Helsinki region's waste management authority are similarly exposed to directives from local politicians. The situation is completely different in a private waste management firm. It can protect itself from the uncertainties of the organizational environment by resolving the pull between waste reduction and disposal as a multi-objective optimization problem. Not surprisingly, the managers of the Finnish branch of one of the largest waste management corporations in the world have no cognitive problems in nailing down the business principles of their operation, which in substance and priority read like a replica of the Ministry of the Environment's policy goals (Lilius 1992).

Conceptually, problem path 1 can be broken with the logic of economic benefit (Figure 6.3). Since excess capacity of waste disposal is the central problem, an optimum needs to be found between waste reduction and disposal. This optimum can be realized by developing ecologically suitable waste management technologies. The need never arises to design excess disposal capacity first and then secure its use with a little help from regulators. Similar conceptual solutions can be developed for the other linear problem paths that the entrepreneurs described (Table 6.5).

CONSEQUENCES OF ENVIRONMENTAL CORPORATISM

The consequences that researchers describe for loop 1 and bureaucrats for loop 3 reflect the belief that serious social disruption would result, should the circular problems be allowed to reinforce themselves without interruption (finding 4). Similar beliefs were observed in the terminal problems of all other interviewee networks as well (Tables 6.1–6.5). The threats that researchers attribute to loop 1 include widespread inability in all sectors of society to think strategically, corporate inability to undertake ecological restructuring, deterioration of ecosystems and permanent change in their functioning, and wars over the environment. Bureaucrats in turn believe that as a result of loop 3, decision making about waste management may become systematically irrational, corporations may completely disregard the principle of sustainable waste management and focus instead on turning a profit by cutting corners in regulatory compliance, and the only role left for environmental regulators may be dumping environmental problems from one official's patch to another.

A second look at waste incineration in the Helsinki region and hazardous waste management at Ekokem illustrates how these threats might materialize. In the Helsinki region's waste management authority, the fusion of implementing and policy making interests within one agency has created a virtual paralysis of far-sighted decision making, in which continuation of the *status quo* is the only common ground that engineers and politicians can find. With no long-term commitment to pursue either waste incineration or waste reduction,

Table 6.5 Summary of the cognitive mapping of entrepreneurs in the Finnish waste management case study

Network element	*Evidence for proposition 1 (conflict between short- and long-term goals)*	*Evidence for proposition 2 (dominance of short-term profit assumptions)*	*Evidence for proposition 3 (environmental corporatism)*	*Evidence for proposition 4 (threat scenarios)*
Problem chain 1: Waste disposal is prioritized over waste recycling and reduction in waste management policy.			The absence of ecologically sound production technologies leads to a prioritization of waste disposal and the design of excess capacity at treatment plants.	
Problem chain 2: Policy choices in waste management are guided by short-term microeconomic considerations.			The emphasis on economic effectiveness and the political inability to take far-sighted decisions leads to waste management problems and diminishing material well-being.	
Problem chain 3: In tough economic times corporations are tempted to cheat to survive.			The lack of clean production technology and economic regulatory instruments in combination with tough economic times leads to corporate cheating.	
Terminal problems resulting from entrepreneurs' problem chains 1 to 3.				Design of excess capacity in waste management technology; reduction in well-being; irrational waste management policy; corporate cheating.

Source: Hukkinen 1995a: 74.

both of which require significant technological and/or institutional changes with lead times in the order of decades, the authority is likely to have plenty of waste but few crisis measures at its disposal when the existing landfills are full. And the policy questions to haunt Ekokem in the long run go something like this: can the Ministry of the Environment take off the owner's hat and put on the regulator's hat without significantly compromising its credibility, when the ageing Ekokem inevitably malfunctions in the next couple of decades? To what extent does the Ministry's emphasis on hazardous waste disposal policy actually impede breakthroughs in low-waste technology?

Today, of course, Ekokem is still one of the exemplary hazardous waste management plants in the world in terms of environmental emissions, and the Helsinki region is far from being flooded by municipal waste. However, it is precisely the absence of acute environmental problems and the high price of acting as if the future mattered that have misled policy makers to opt for the *status quo* as the best bet to avoid future problems.

The principles for the design of institutions that would make preparing for the long-term future an attractive option for the decision maker are fairly clear. Environmental corporatism in Finland should be replaced with an institutional framework that supports the sustainability goals held widely by both policy makers (as the interviews indicate) and society in general (as Uusitalo 1991 has shown). Without such support, the sustainability goals are not likely to be openly explained, debated, settled and acted upon within the parliamentary decision making system. The unifying principle should be the administrative and procedural separation of conflicting environmental policy interests, and the simultaneous empowering of sustainability goals with appropriate institutions. In the terminology of institutional economics, organizational uncoupling of conflicting interests in waste management lowers the price individual policy makers have to pay for their actions (North 1992). This would provide the freedom for individuals to incorporate their ideas and ideologies first into the operating assumptions and eventually the choices they make, and thus promote an open political debate and settlement of the issues.

In Finnish waste management, conflicting interests are prominent at all three levels of institutional rule. At the constitutive level, environmental corporatism ensures that short-term economic issues get priority over longer-term environmental ones. At the collective choice level, the organizational fusion of waste reduction and disposal puts long-term waste reduction policy at a disadvantage when considered side by side with the urgent task of waste disposal. And since the operational rule holds environmental regulation as a whole an inseparable part of the day-to-day operations of environmental management, long-term regulatory policy does not exist. Under such circumstances, the seeds of institutional autonomy for long-term interests could be sown by establishing an independent agency with the sole responsibility of waste reduction and a political and/or regulatory body with an agenda to make long-term environmental policy. I will elaborate the specifics of these recommendations in Chapter 8.

It should be emphasized that the principle of separating conflicting policy constituencies does not aim to polarize environmental conflicts between sustainability and profit maximization, but rather to resolve issues through the efficient conflict resolution mechanisms that already exist in Finnish society. Where such mechanisms do not exist, they should be created.

WHEN THE ABSENCE OF CONFLICT IS THE PROBLEM

This case study has explored obstacles to ecological sustainability in a corporatist state where social conflict has been institutionalized by including all interested parties in negotiations. Scandinavian countries are often categorized as non-exclusively corporatist. The other, exclusive type of corporatist state institutionalizes partnership and consensus, but exclusively among parties considered legitimate representatives of social interests. The Netherlands is the prototype of this industrial order (Therborn 1992). How have policies aiming at ecological sustainability fared in exclusive corporatism?

Inferring from the principles of exclusive corporatism, one would expect it either to incorporate ecological aspects fully in the consensus-building, should ecology be regarded as a legitimate corporatist party, or effectively to exclude them from the consensus, should ecology not be accepted as a corporatist player. According to Opschoor and van der Straaten (1993), the latter is the case in the Netherlands. Despite impressive policy statements about ecological sustainability, as expressed in the National Environmental Policy Plan (NEPP) of 1989 (Ministry of Housing, Physical Planning and the Environment 1989) and its later version NEPP 2 (Ministry of Housing, Spatial Planning and the Environment 1994), environmental policy implementation has contradicted its stated objectives. NEPP does not rely on new, economic instruments for environmental policy, but rather assumes that clean technologies will disseminate through the institution of covenants, i.e. voluntary agreements on environmental management between the government and the chief industrial sectors. As such, covenants are a logical extension of consensual corporatism to the environmental policy sector. Although Opschoor and van der Straaten stop short of mentioning corporatism, their conclusion of the Dutch situation resembles closely the one presented here. According to them, labour and capital have been able to build up such a strong position in the state machinery that they can 'solve' their struggle by using up the 'powerless' production factor, i.e. natural resources.

At issue here are trade-offs between labour, capital and nature (Pearce and

Turner 1990). In principle, policies could be directed toward ecological sustainability by developing ways of including ecology in corporatist negotiations and agreements. In practice, however, this is impossible due to the very logic of corporatism. Even the most inclusive corporatist societies, such as Finland, effectively prevent ecological sustainability from ever entering as a party in corporatist negotiations, because decision makers themselves conceptualize environmental issues in terms of unproblematic operating assumptions. As a result, the central long-term environmental policy problem in Finland is not overt environmental conflict, but rather its absence. Policy makers are cognizant of ecological sustainability, but only in a dissonant and unresolved way, as the circular arguments of the interviews illustrate.

Since the exclusion of ecology from corporatist deals takes place at a very fundamental, cognitive level, the question is not one of balance of power in the state machinery, as Opschoor and van der Straaten (1993) argue. Reshuffling power among existing corporatist interest groups would in no way empower ecology and natural resources in corporatist negotiations. The dismantling of the corporatist decision making structures and procedures is required before the latent ecological awareness among policy makers will materialize in ecologically sustainable policies. Note that this is not meant categorically to disallow the popular prescription to 'include all stakeholders in the policy process'. The aim here is rather to draw attention to the dangers of structural confusion, be it in the policy process or in the organizations participating in that process, of environmental interests that are in profound conflict.

Environmental corporatism is an issue not just in Finland and the Netherlands. Most European nation states have developed some form of corporatism (Evans *et al.* 1985; Pekkarinen *et al.* 1992; Wilson 1989). The European Union's decision making institutions have largely inherited these characteristics (Duff *et al.* 1994; Harrop 1989; Nugent 1989). Environmental corporatism is therefore a concern for the EU's environmental policy making institutions. I will discuss how these concerns should be taken up in the design of EU's environmental institutions in Chapter 10.

The next chapter presents the last case study, which focuses on environmental management in China. The case study is presented to broaden the geographical coverage of the empirical analyses, and more importantly, to show that our approach can be applied in a cultural and institutional setting different from that of the three earlier case studies.

7

ENVIRONMENTAL MANAGEMENT IN CHINA

The case studies presented in the preceding chapters come from the Western hemisphere. We now move to China, where the culture, institutional setting and stage of development differ considerably from the earlier case studies. The findings of the Chinese case study largely support the main proposition of *Institutions in Environmental Management* concerning the mutually reinforcing relationship between environmental institutions and the cognitive models of decision makers. Fundamental differences in the institutional setting and the stage of social and economic development, however, became evident early in the data collection phase of the research. Because of cultural and language barriers, caution is necessary in drawing inferences from the interviews. As a result, the following investigation and the institutional design implications drawn from it rely much more than the previous ones on an analysis of the conceptual and cultural context of Chinese environmental management. Despite the reservations and limitations, China brings valuable nuances to our understanding of the feedback between mental models and environmental institutions. It is also a country one cannot omit from the analysis, if simply because of the magnitude of its environmental problems.

China is rushing toward prosperity. But at a cost. Its economy grew at the stellar pace of almost 10 per cent per year since economic reforms began in 1978 (Rohwer 1992). While the GDP per capita was a modest 370 US dollars in 1990, the GDP adjusted to purchasing-power parity was 1,950 US dollars, making the Chinese economy roughly the same size as that of Germany (Cottrell 1995). The dramatic increase in material wealth has taken its environmental toll. In 1993, more than 300 Chinese cities were short of water, with 100 in acute distress. During the past three decades, about 15 million hectares of arable land were converted to other uses, including industrial and urban. Timber scarcity threatens both forestry and biodiversity. Coal provides 76 per cent of primary energy in China, contributing to some of the world's highest urban sulphur and particulate readings, and making lung diseases the leading cause of death in the country (*Economist* 1992; Ryan and Flavin 1995). The extent and intensity of China's environmental degradation have become, according to Smil (1993), critical determinants of the nation's development aspirations.

Such aspirations China certainly has. Despite the rapid economic development, it still is a poor and populous country. Of her 1.13 billion people, eighty million were estimated to be poor in 1993. Although the growth rate of population has slowed considerably since the 1980s, when the government adopted a policy of one child per couple, population will in absolute terms grow well after 2020, because the population at child-bearing age will not peak until the 1990s or later. The relatively low farmland area per capita seriously limits efforts to increase food production. Chinese leaders consider the transfer of population from rural areas to non-agricultural employment to be a precondition for the rational use of farmland, in other words, the feeding of the population. Urbanization and economic development, however, have brought with them the problems of urban encroachment on precious agricultural land and dietary change to animal proteins. The production of animal proteins has in turn pushed grain consumption up (Christiansen and Rai 1996; Ryan and Flavin 1995).

Hunan Province of China, where this case study took place, mirrors the problems of the nation. It is located at the meeting point of an agricultural past and a rapidly industrializing future. I interviewed environmental managers in the capital of Hunan Province, Changsha, which lies about 700 kilometres northwest of Hong Kong. Hunan used to be an agricultural region, but has over the past decades developed into a mix of agricultural and industrial enterprises. The main industries produce non-ferrous metals and steel, paper, construction materials and textiles. According to the interviews, the industries have brought not just wealth but also many environmental problems, such as sulphur dioxide emissions from coal burning, heavy metal emissions from the non-ferrous metal smelters, waste water pollution from numerous small paper mills, construction dust from rapid urban growth, and solid waste and noise problems in cities (interviews 4, 5 and 6; workshop).

As environmental problems grow more serious in China, Western entrepreneurs will see increasing business opportunities in environmental management. Evidence from Western industrialized countries shows, however, that designing and implementing environmentally sound industrial production systems is not just a question of technological know-how and equipment. Inadequate attention to the institutional and organizational issues of environmental management often undermines, if not invalidates, resource commitments in environmental technology and engineering. Western corporate environmental management has therefore striven to be a proactive, strategic and autonomous activity by a firm to achieve competitive advantage. It devotes more resources to preventing pollution with clean technologies rather than controlling pollution from existing production processes with end-of-pipe technologies. The underlying rationale is to maintain the ecological basis of industrial and economic activity (Beaumont *et al.* 1993; Buchholz 1993; Harrison 1993; Hukkinen 1995b; Welford 1993; Welford and Gouldson 1993).

This chapter's main finding is that corporate environmental management as conceptualized in the West does not yet exist in China. Many town and village enterprises (TVEs) see environmental management as no concern of theirs, while state-owned enterprises (SOEs) treat it primarily as a technological challenge to control pollution from existing production processes. For the SOEs, environmental management is a tactical reaction to environmental regulations stipulated by the state bureaucracy, which firms largely depend on for financial support. For both TVEs and SOEs, environmental management is undertaken only if the short-term economic profitability of a firm is not threatened.

My study of environmental management in Hunan Province began as an attempt to lay the groundwork for cooperation between Western and Chinese partners in areas of environmental management. The study was part of the Sino-Dutch Management Training Project (SDMTP), which the Dutch development cooperation agency launched in 1987 with the Hunan Economic Management College (HEMC) in Changsha. The original aim of the environment component of the project was to establish permanent environmental management training at HEMC and to facilitate environmental business links between European and Chinese companies. However, in the course of the work it became evident that fundamental differences between China and the West exist in the way environmental management is conceptualized and put into operation. This study outlines what those differences are and how they should be addressed in future environmental management cooperation between China and the West.

I have written the chapter very much the way my exploration of Chinese environmental management unfolded during two visits to Hunan Province, first for two weeks in November 1994 and then four weeks in April 1995. Efforts to sort out the fundamental conceptual differences led the way to a more detailed account of the institutional, administrative and enterprise-level activities in environmental management. The chapter concludes with a sketch of some common points of departure for Chinese and Western counterparts in developing corporate environmental management in China. Before entering a detailed description of the interviews, however, I want to recount two experiences that illuminate the conceptual differences one has to be aware of when attempting to understand environmental management in a culture fundamentally different from Western ones.

CONCEPTUAL DIFFERENCES

My first Chinese environmental management conundrum was a meeting with a group of managers from the Zhuzhou Non-ferrous Metals Factory during the November 1994 visit to Hunan. I asked my hosts to describe some of the environmental management challenges they were facing. In response, they either detailed challenges in the design and implementation of the environmental protection plan they had prepared in cooperation with the environmental regulator, the Hunan Province Environmental Protection Bureau; or they provided dust particle counts in their air emissions and countered the question with a request for further information on the latest Western technology in the removal of dust particles from smelter air emissions (interview 3). These were answers to the question I never asked, namely what other environmental problems, apart from those relating specifically to enterprise-level environmental management, were they facing. I left the meeting with the impression that what Chinese SOE managers called environmental management was in my mind environmental regulation and engineering.

The second enigmatic encounter took place during my visit to Changsha in April 1995. One of my tasks was to design the curriculum for an environmental management course to be arranged at HEMC. My counterpart there was LM, who taught industrial economic management and was going to be responsible for teaching a future environmental management course. LM was already devoting some time to industrial environmental management in her course. I asked LM to describe the details of her class discussion on environmental management.

LM: I teach environmental strategy.
JH: And what do you talk about there?
LM: The National Environmental Protection Plan of China.
JH: Ah, so that's really more like national environmental policy.
LM: But it is the same thing in China!

Remember, LM's students were practising enterprise managers. In further discussions, I confirmed that in her mind environmental management, policy and strategy were inseparable, one and the same thing at national, provincial and municipal levels, both in government and in enterprises.

Confused by the apparent differences between what I understood to be environmental management and what my Chinese contacts did, I decided to consult a dictionary. Having no knowledge of Mandarin, I first asked my interpreters to write down the Mandarin translations they had used in the interviews for the English environmental terms I used. They then looked up the English translation for the Mandarin words in a Mandarin–English dictionary. Differences between the English definition of the term I had originally used and the English translation of the Mandarin term my interpreters had used might give an indication of conceptual discrepancies between English and Mandarin environmental terminology (barring, of

Table 7.1 Comparison of English and Mandarin terminology in environmental management

English term used by JH	*Mandarin term used by interpreters*	*Mandarin–English dictionary*	*American Heritage Dictionary*
management	Guan Li	control; supervise; command; run; administer; handle	contriving; arranging; getting along
policy	Zheng Che	regulation	prudence; shrewdness; care and skill in managing one's affairs or advancing one's interests
technology	Ji Shu	skill; technique	application of science in industry and commerce

Sources: Chinese–English Dictionary 1985; Chinese Management Science Dictionary 1985; Modern Chinese–English Dictionary 1988; American Heritage Dictionary 1986.

course, the chance that the interpreters simply had not used the correct translation). Table 7.1 shows partial results of this exercise.

Table 7.1 points out translation differences of the terms 'management', 'policy' and 'technology'. 'Management' appears to have no other connotations in Mandarin apart from pure command and control. In English, it also refers to 'contriving', 'arranging' and 'getting along'. Just as in Mandarin, 'policy' has the regulatory connotation in English, but also refers to 'prudence', 'shrewdness' and 'care and skill in managing one's affairs or advancing one's interests'. 'Technology' appears to be understood more comprehensively in English than in Mandarin, comprising the 'application of science in industry and commerce'. One should, of course, be careful in terminology comparisons such as the above when the two languages are so fundamentally different and changing. English translations in a Mandarin–English dictionary are very limited and do not cover the full range of meanings attributable to a given Mandarin concept.

The dictionary comparisons do, however, support the conceptual discrepancies that emerged in discussions with Chinese counterparts and interviewees. Interviews revealed the dual meaning Chinese managers and educators give to environmental management, one referring to the micro-level control of environmental problems through engineering applications, the other to the macro-level control through government regulation (interviews 3 and 6; workshop). Both would also qualify as textbook examples of command and control management, containing not even a hint of the more conventional notions of management, such as horizontal partnership, networking, negotiation and persuasion. Instead, LM's comment about corporate environmental strategy being equal to national environmental policy planning reflects, I believe, a not-uncommon regulatory and control-oriented conception of environmental management. Finally, when asked about environmental management problems in their enterprise, the managers in interviews 1, 2 and 3 answered in very specific engineering concepts. In sum, the challenge in their minds was not one of applying science to solve a management problem but of finding an engineering device to fix a specific technical problem.

The conceptual differences correspond to two characteristics which, according to analysts of China, run deep there: respect for authority and pragmatism. The millennia of dynastic rule have made respect for authority the norm of politics, supported by the strict hierarchical control required by irrigation technology and the authoritarian philosophy espoused by Confucius (Christiansen and Rai 1996; Kristof and WuDunn 1994; Wittfogel 1957). Respect for authority is said to permeate Chinese consciousness to the extent that modern mainland Chinese have been found to equate ethical conduct with adherence to laws and regulations (Szalay *et al.* 1994). The only effective avenue for expressing individual creativity appears to have been the invention of clever techniques for solving the practical problems of daily life.

Thus, Chinese cultural inheritance leaves less room for environmental management as understood in modern industrialized countries. Environmental management as an autonomous, strategic activity that the enterprise undertakes to influence and adapt to emerging environmental situations is the 'missing middle' in a Chinese firm's dealings with ecological questions. In SOEs it is pushed aside by the overriding concern to satisfy the demands of the superior in a chain of command, i.e. the state's environmental protection officials. The organizational aspects of corporate environmental management are thus reduced to authoritarian politics and the only autonomous field of expertise left for SOE managers is that of nuts-and-bolts environmental engineering. As will become clear, the other corporate approach to environmental challenges is the one adopted by many TVEs, namely to forget about the environment altogether.

TALKING TO ENVIRONMENTAL MANAGERS IN HUNAN PROVINCE

To obtain a more detailed picture of the environmental management problems facing the Changsha region, I conducted six interviews and a workshop with chief environmental policy makers in provincial government and industry. Again, the original objective of the interviews and the workshop was to develop environmental management training in the region and facilitate links between European and Hunanese environmental management operators. While falling short of this goal, the interviews did reveal insights on the institutions and organizations of environmental management in Chinese enterprises.

The first contact with the region's environmental management organizations was established in a workshop arranged in November 1994 in collaboration with the Hunan Province Environmental Protection Industry Association, a group designed to promote environmentally sound production and facilitate collaboration between enterprises and environmental regulators. The workshop produced a list of eight agencies and enterprises that the participants felt would provide an exemplary snapshot of the environmental management situation in the Changsha region. Two of the organizations, the Hunan Province Environmental Protection Bureau (the provincial regulator) and a paper mill, could not be interviewed due to scheduling and logistical problems. The list of interviewed organizations is in Table 7.2, together with summary data on their environmental management functions.

Although few, the interviewees represent the main players in Changsha region's environmental management sector, including industrial enterprises, environmental regulators and an environmental research institute. The enterprises range from small (such as interviewee 2, Hunan Lichen Industrial Corporation) to very large (such as interviewee 1, the Xian Tan textile factory, which is the largest of its kind in Hunan Province) and represent key industrial sectors in the province. However, the interviewed firms are all SOEs, and the absence of TVEs, which are currently responsible for most of China's economic growth and environmental problems, is a major blindspot in the sample (Table 7.2).

The interview format differed somewhat from the previous case studies. Four of the six interviews were with a group in which anywhere between two to twenty environmental experts from the organization were present at the same time. The group format was specifically demanded by those interviewed, because they felt it would facilitate a comprehensive exchange of environmental management information. Furthermore, the problem descriptions are based on what the interpreters claimed the interviewees said. Both factors may have produced information different from what I would have obtained had the discussions taken place in Mandarin with individual experts.

As one would expect on the basis of the earlier discussion, the interviewees gave a wealth of detailed accounts of the environmental engineering problems that managers in government agencies and industrial enterprises are facing. When the purely technical problems were left out, fifteen different problem statements dealing specifically with environmental policy and regulation were identified in the six interview groups. The perceived problems had to do with TVEs purportedly careless about environmental management, lack of money to invest in environmental improvements, enterprises reluctant to change their management practices, government

Table 7.2 Data on firms and agencies interviewed on environmental management in China

Firm or agency	*Main functions or products*	*Number of employees*	*Main pollutants of concern*	*Amount of waste water generated (m^3/d)*
Interview 1 Xian Tan Textile and Printing Factory	Cotton and linen clothes and textiles	11,000	Printing waste water with high levels of organics	5,000
Interview 2 Hunan Lichen Industrial Corporation	Soaps, detergents, toothpaste	200	Organics in waste water; sulphur dioxide from coal burning; volatile organics from package printing	50
Interview 3 Zhuzhou Nonferrous Metals Factory	Lead, zinc, copper and other nonferrous metals	8,000	Lead fumes and dust; sulphur dioxide from coal burning; waste slag and clinker	20,000
Interview 4 Changsha City Environmental Protection Bureau	Regulation and management of Changsha city environmental protection	200	N/A	N/A
Interview 5 Hunan Province Environmental Protection Research Institute	Environmental research and assessment; formulation of government environmental policy	150	N/A	N/A
Interview 6. Hunan Province Environmental Protection Industry Association	Promotion of environmentally sound industries under the auspices of Hunan Province Environmental Protection Bureau	(106 member firms in 1994)	N/A	N/A

Note

In addition, a workshop was held on 15 November 1995 with the following firms or agencies: Hunan Province Environmental Protection Bureau (3 participants); Hunan Province Environmental Protection Industry Association (2 participants); Hunan Province Environmental Protection Research Institute (2 participants); Zhuzhou Nonferrous Metals Factory (1 participant); Miluo Yongqing Environmental Protection Equipment Factory (air pollution control equipment) (1 participant); Changsha Huade Industry Limited Corporation (automobile air emission control equipment) (1 participant).

Table 7.3 Hunan Province environmental management problems organized according to topic

Interview	*TVEs do not care*	*No money*	*Stuck in old ways*	*Economy 1st, environment 2nd*	*Hindrances to new directions*
1 Xian Tan	1.1 Regulators are unable to collect pollution fees from the TVEs, who generally do not care about environmental management.	13.2 Since China is not a very rich country, there is not much money for environmental investments.	1.3 Since enterprises that pay the pollution fee also get a low-interest loan for environmental improvements, good environmental performers end up paying for environmental protection.	15.4 Economic development always wins over environmental protection in government policy, because Deng has said 'Economy first, environment second'.	1.5 Although government policy puts pollution prevention before pollution control, enterprises cannot get environmental improvement loans for clean technology investments.
2 Lichen		24.2 Environmental standards are difficult to meet in enterprises, because they have little money available for environmental improvement investments.			
3 Zhuzhou		13.2 Since China is not a very rich country, there is not much money for environmental investments.			3.5 Employees need environmental protection training.

Table 7.3 (Continued)

Interview	*TVEs do not care*	*No money*	*Stuck in old ways*	*Economy 1st, environment 2nd*	*Hindrances to new directions*
4 City		24.2 Environmental standards are difficult to meet in enterprises, because they have little money available for environmental improvement investments.	45.3 Regulators are not strong enough to confront enterprises and economic policy makers, because many people would become unemployed if regulations were to be enforced.	4.4 The city's environmental problems are due to its rapid economic and industrial development.	
5 Research	56.1 The polluting TVEs are many, small and short-lived, which places them outside the control of environmental regulators.		45.3 Regulators are not strong enough to confront enterprises and economic policy makers, because many people would become unemployed if regulations were to be enforced.	15.4 Economic development always wins over environmental protection in government policy, because Deng has said 'Economy first, environment second'.	5.5 The number of small TVEs increases rapidly, which makes it difficult for them to merge into larger enterprises that could more easily invest in environmental protection.
6 Association	56.1 The polluting TVEs are many, small and short-lived, which places them outside the control of environmental regulators.	6.2 Subsidies to old and polluting SOEs absorb government funds from environmental management.	6.3 SOEs are old, large, very inefficient and polluting, but cannot be allowed to go bankrupt because they serve an important social function.	6.4 TVEs can always find ways around environmental policy because the regulatory system is weak.	6.5 Government policy is to transfer industrial production from the old and polluting SOEs in the cities to TVEs in the countryside.

policy prioritizing the economy over the environment, and institutional and policy obstacles standing in the way of environmental improvements (Table 7.3).

The interviewees placed the problem statements in causal relationships with each other. When the problem networks obtained from the six interview groups were aggregated, network configurations similar to those observed in the Californian and Finnish case studies emerged. Two initial problem paths led to a loop, which in turn was seen as giving rise to three terminal problems (Figures 7.1 and 7.2). The following discussion will first describe the loop and then the network surrounding the loop.

The problem loop (Figure 7.1) explains how local environmental regulators find themselves in a weak position with respect to economic policy makers, because forceful regulation might cause widespread unemployment (problem statement 45.3). The regulators find such an outcome potentially threatening to themselves, particularly since the highest levels of government prioritize economic development over the environment (as underlined by problem statements surrounding the loop, to be discussed shortly). The weak regulatory system is perceived to allow the TVEs to disregard environmental controls completely (problem statement 6.4). As a result, the uncontrolled TVEs proliferate and boost the economy (problem statement

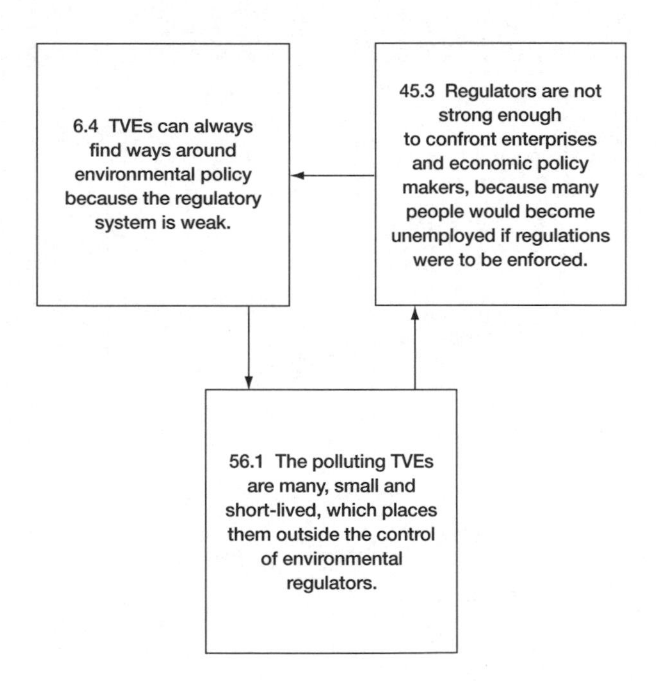

Figure 7.1 Loop described by Chinese environmental experts.

56.1), which further strengthens the position of economic policy makers and weakens the position of environmental regulators.

As in the Californian and Finnish case studies, the loop is held together by a goal conflict. On the one hand, the loop reflects a deep concern over the economic hardships that vigorously implemented environmental regulation might bring with it. On the other hand, the loop reflects at least equally grave concern over environmental degradation, which is seen inevitably to result from having the TVEs act on their own as the engines of China's economic propulsion. But, in contrast to what was observed in the earlier cases, the loop, as it is, cannot be said to reflect *cognitive* dissonance, because it is formed from the aggregation of several individuals belonging to different interest groups. As such, the loop would appear to reflect inter-organizational, not intra-personal, conflict.

On closer examination of the problem statements, however, an argument can be made for at least a potential for cognitive dissonance among the Chinese experts. First, at least seven problem statements (1.1, 13.2, 15.4, 24.2, 4.4, 56.1, 6.4 in Table 7.3) contain polarized arguments by frustrated environmental managers who feel they have been overruled by economic policy makers, thus indicating that the economy versus the environment is a defining axis of the Chinese environmental debate. Furthermore, three problem statements clearly express a pull between conflicting policy goals at the individual level. Problem statement 45.3 in Figure 7.1 shows how the wish of an environmental manager to confront economic policy makers is compromised by the realization that environmental regulation might force many people out of jobs. Similar dissonance is evident in problem statement 6.3 in Figure 7.2 and a related problem statement 6.2 in Table 7.3, which reflect a pull between the goal to shut down inefficient, subsidized and polluting SOEs, and the operating assumption that the shutdown is infeasible, because SOEs provide important social services. These three problem statements have exactly the same logical structure as the issues of the Colorado case study and the operating assumptions of the Californian case study. Accordingly, the statements will here be treated as indicators of a potential cognitive dissonance. Proof of cognitive dissonance will have to wait, as only a few individuals mentioned such issues, the interview sample was small, and no structural evidence (i.e. individual circular argumentation) could be found for cognitive dissonance. That said, analysis of the institutional, managerial and regulatory context will provide additional support for the proposition that cognitive dissonance is a reality to be reckoned with in Chinese environmental management.

Two initial problem paths lead to the problem loop, which in turn gives rise to three terminal problems (Figure 7.2). The initial problem paths in Figure 7.2 shed further light on the problem of proliferating TVEs mentioned in the loop. Since SOEs are perceived to be inefficient and polluting, the policy is to transfer production from them to TVEs in the countryside,

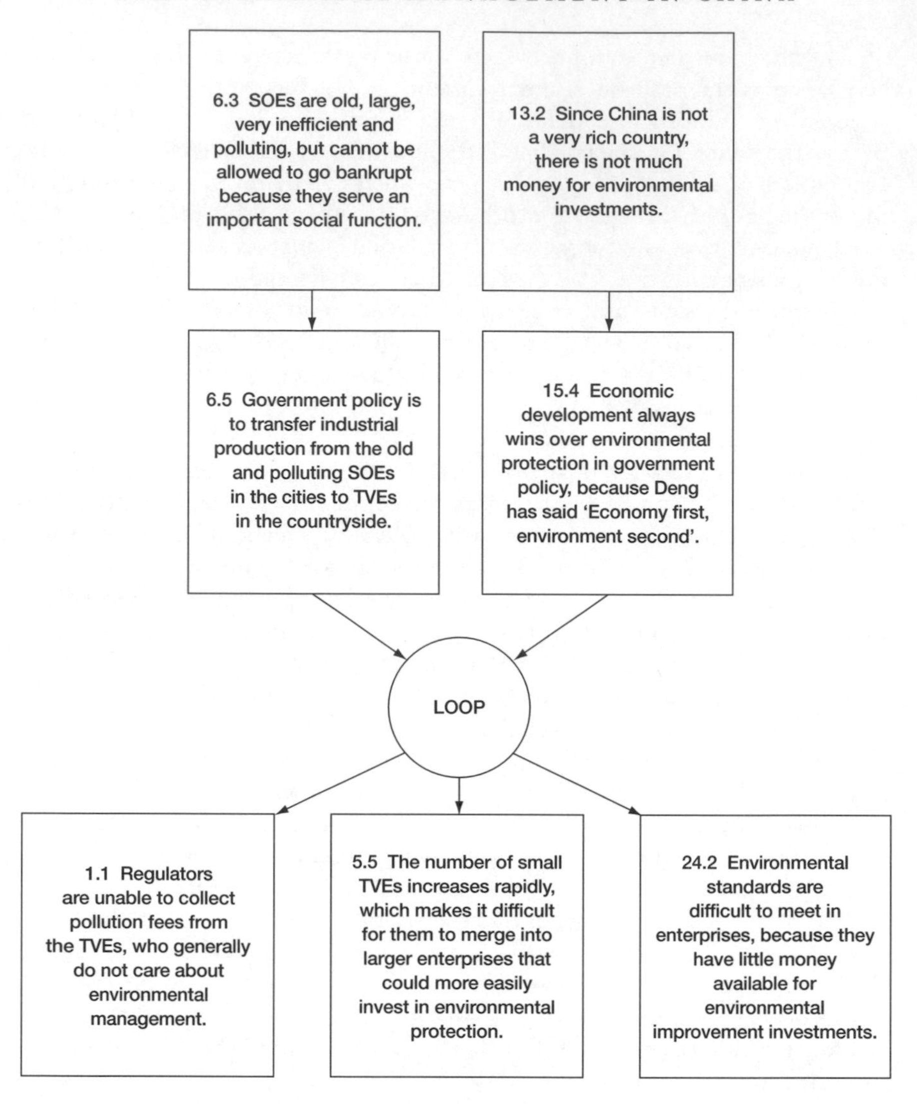

Figure 7.2 Network surrounding the loop by Chinese environmental experts.

which only reinforces the relative strength of the TVE-sector (problem statements 6.3 and 6.5). Since China is poor and needs economic development, TVEs get additional support from the highest levels of Chinese economic policy making (problem statements 13.2 and 15.4). The terminal problems describe the end results of the initial paths and the loop. Environmental management is feared to lose its economic foundation, because the regulators are unable to collect pollution fees from the enterprises (problem statement 1.1). TVEs are expected to proliferate even faster in the future, which will prevent them from ever fully consolidating into larger enterprises that could more easily pool resources for pollution control (problem statement 5.5). Finally, the state of the environment is expected inevitably to worsen, because enterprises cannot meet environmental standards (problem statement 24.2).

The results of the interview analysis can be summarized succinctly. First, the interviews show clear signs of a conflict of interest between environmental regulators and economic planners in China. Second, the interviews contain evidence that environmental policy makers and regulators have internalized the conflict as cognitive dissonance. They think environmental regulations should be vigorously enforced, but also recognize that such action is infeasible because it might jeopardize China's rapid economic growth. Finally, the interviews display the concern environmental managers have over the grave results of allowing the current imbalance between environmental and economic policy to continue. The next two sections reinforce these findings by showing that environmental policy makers and regulators have good reasons for harbouring cognitively dissonant mental models.

INSTITUTIONS OF ENVIRONMENTAL PROTECTION

Two central factors affecting corporate environmental management in any country are the types of social rules (i.e. institutions) that have evolved to guide environmental management, and how the power to apply the rules has been organized (i.e. the administration of environmental management). According to this study, the institutions that govern environmental management in China are characterized by hierarchical administration extending from central to local government, combined with central planning and authoritarian control based on laws, regulations and traditions.

China's environmental protection administration has a strictly hierarchical structure. The National Environmental Protection Agency (NEPA) is the highest agency with comprehensive responsibilities in environmental protection, including the implementation of national environmental policy, law and regulation; the design of regulations and standards; long-term planning; organizing environmental monitoring; supervision of environmental research and education; and international cooperation and communication. The hierarchy extends to the provincial level. In Hunan Province, for example, the Hunan Province Environmental Protection Bureau has roughly the same responsibilities as NEPA, but at the provincial level. The bureau has departments with responsibilities, such as Communist Party relations, financial affairs, planning, education, research, monitoring, media and relations with industry and the scientific community (Dai and Zhang 1984; *Environment Yearbook of China* 1993).

The laws, regulations and policies that environmental authorities administer are enforced largely through command and control principles. Novel regulatory approaches, such as economic instruments and agreements between government and industry, are at most discussed in professional journals (Liu 1994). NEPA's list of its key achievements in China's environmental protection in 1994, for example, includes the incorporation of the decision of the Central Committee of the Communist Party on the building of a socialist market economy, speeding up environmental legislation and enforcing the law strictly, and completing environmental regulations and management systems (National Environmental Protection Agency 1994b). Sanctions for non-compliance include penalty fees, which sometimes take characteristically Chinese forms. In 1993, for example, the Environmental Protection Committee of the National People's Congress, the Central Propaganda Ministry, the Broadcasting and Television Ministry, and NEPA launched 'China's Environmental Protection Mission of the Century'. During the mission, the agencies reported and gave wide publicity to environmental problems and illegal actions they found in nineteen provinces (National Environmental Protection Agency 1994a). Similar publicity is used at the local level. In Hunan Province, chief executives of enterprises found responsible for significant environmental accidents have been criticized publicly on television (interview 3). This form of punishment is probably more severe than it appears from the Western perspective, given the sensitivity of mainland Chinese to public humiliation and loss of face (Szalay *et al.* 1994).

Environmental management *system* is a key term in NEPA's list of achievements. It also emerged in discussions with government officials and enterprise managers in Hunan Province. To them, environmental management system meant activities ranging from environmental protection planning and policy setting by NEPA, to environmental regulation and standard setting by provincial government, to monitoring by city government, to technology design and operation by the SOE (interview 6; workshop). From the government point of view, SOEs are thus organs of the state, there to implement government policy. Separation between regulatory and implementing powers, which is one of the centrepieces in the blueprints – although not always the reality, as the earlier cases have shown – of environmental institutions in most Western democracies, is virtually non-existent in China. Conflicts of interest in the Western sense are perceived to be impossible, because the corporate environmental management machinery working toward China's common environmental protection goal includes NEPA, SOEs and all levels of environmental regulation between them. The exception to this system are the TVEs.

ENVIRONMENTAL PROTECTION IN ENTERPRISES, INCLUDING TVES

Hierarchical administration and authoritarian regulation leave just two avenues for the SOE to influence its own environmental protection work. One is to manoeuvre tactically with respect to government regulations. The other is to develop and install pollution control technologies. What is missing between the two approaches are environmental management measures of the kind currently preoccupying the minds of Western corporate strategists.

The combination of environmental standards and pollution fees intended to encourage regulatory compliance in China has had the unfortunate effect of making the good environmental performers carry the burden of pollution control costs (problem statement 1.3 in Table 7.3). All enterprises are supposed to pay an annual 'standard fee' as long as their emissions stay below the environmental standard set by the regulators (Figure 7.3). However, if the

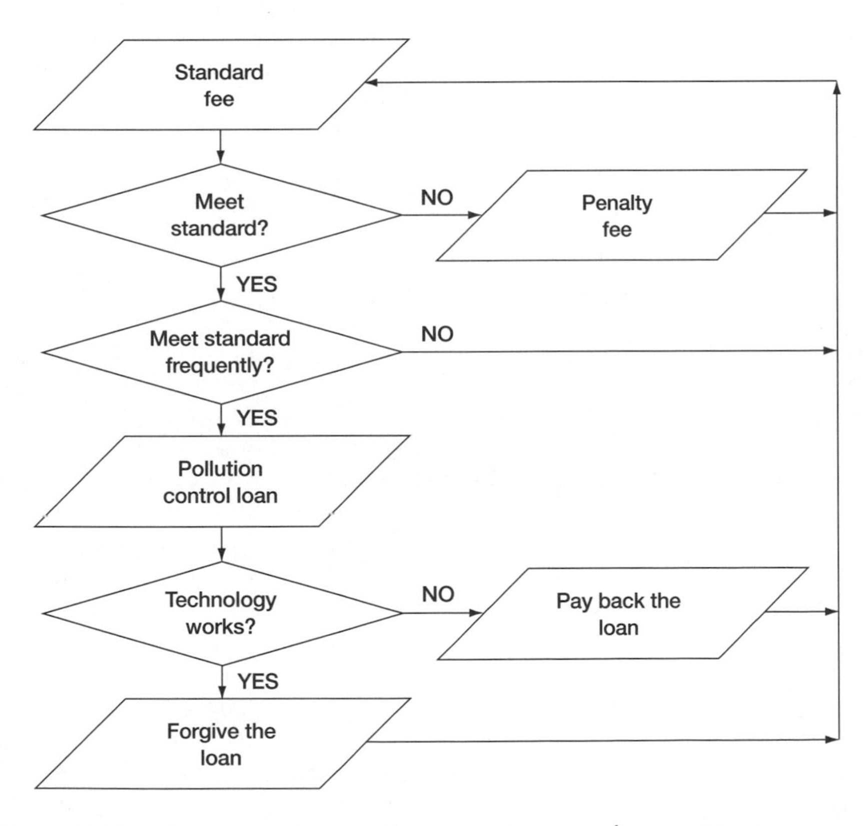

Figure 7.3 Regulatory sanctions in Chinese environmental management.

enterprise exceeds the emission standard, it should pay a 'punishment fee' in addition to the standard fee. Those firms that pay the annual standard fee are eligible for a low-interest loan from the provincial environmental protection bureau for investments in pollution control technology. If the technology operates well, the firm may be forgiven the loan altogether (interviews 1, 4 and 5). In practice, the arrangement has divided firms into non-paying polluters and paying non-polluters. The non-paying polluters are largely TVEs, which are often out of reach of regulatory control. The paying non-polluters are those environmentally aware SOEs that want to ensure a return on their annual pollution fee by continuously investing in pollution control technology and thus ensuring they do not have to pay back the government loan. As one environmental manager put it, quoting a Chinese proverb, 'Hair of goat comes from goat' ('Yang Mao Chu Zai Yang Sheng Shang'; i.e. the farmer benefits from keeping his goat happy, just as state enterprise benefits from keeping the government happy by investing in pollution control) (interview 1).

The pollution fee system has also directed environmental investment to end-of-pipe pollution control technology instead of clean technology focusing on process improvements. The provincial environmental protection bureau does not award low-interest loans for production process improvements – even if the factory's environmental performance would improve – because they are considered normal business investments (problem statement 1.5 in Table 7.3). Not surprisingly, Chinese environmental protection analysts do acknowledge the benefits of clean production, but point out that in practice the emphasis of environmental protection at the enterprise level is on end-of-pipe pollution control technology (Yang 1994).

Completely missing between environmental regulation by the state and pollution control engineering by the enterprise is the set of strategic activities, such as environmental audit, environmental impact assessment, life cycle assessment, environmental accounting and environmental marketing, with which modern management literature defines corporate environmental management and industrial ecology. In contrast, the Chinese SOE views itself as an implementor of the National Environmental Protection Plan by means of pollution control engineering.

TVEs are an altogether different matter. By and large, interviews conducted during this study indicate that TVEs are considered the villains of environmental protection in China, at least from the perspective of informed outsiders (Table 7.3). Many in number, small in size, short-lived and geographically scattered throughout the countryside, they are effectively beyond the control of environmental protection officials (Ryan and Flavin 1995). The official publication on industrial environmental pollution in China can only provide an estimate of the contribution of TVEs to the overall pollution burden from industry, putting it in the range of 13–15 per cent of total industry pollution (Liu 1994).

The success of TVEs in avoiding environmental protection is directly related to the Chinese government's economic policies. While the privately and collectively owned TVEs are the economic success story of the past fifteen years, 60 per cent of SOEs lost money in 1993. However, since the SOEs provide not just jobs, but also housing, education and health care for their workers, government policy views them as the backbone of China's national economy (Cottrell 1995). As a result, the government subsidies that SOEs receive also guarantee the persistence of an old and polluting industrial infrastructure. TVEs, on the other hand, operate beyond government influence, including environmental subsidies and regulation. Theirs are the slash-and-burn tactics of operating for a year in one rural village with little regard for the environment, going bankrupt the next year, only to be reborn in some new ownership configuration the year after (problem statements 56.1, 6.4 and 6.5 in Table 7.3). Rural China offers TVEs a solid base for operation. The rural society whose economies TVEs have primarily been boosting has a long-running tradition of successful avoidance of government control (Szalay *et al.* 1994).

THE AUTHORITARIAN CHALLENGE TO ENVIRONMENTAL MANAGEMENT

The central proposition of this book is that the actions of environmental managers and regulators are largely determined by a feedback mechanism between the institutions within which they operate and the mental models with which they understand the environmental challenges they are facing. More specifically, the predominant institutional order in the case study societies prioritizes short-term economic concerns over long-term environmental ones, which persuades environmental decision makers to adopt cognitively dissonant mental models. On the one hand, they think that long-term environmental concerns should guide environmental decision making. On the other hand, they believe that short-term economic concerns will in the end determine which policies will be implemented.

The case study on Chinese environmental management largely supports the proposition. Interviews with Chinese environmental managers and regulators give strong indications of cognitively dissonant mental models. Environmental decision makers would like to enforce environmental regulations that take into account long-term environmental sustainability, but do not do so because of their weak institutional position with respect to economic policy makers, and their fear that regulation might trigger mass unemployment.

Environmental decision makers' perception of the institutional constraints is correct. Analysis of the institutional and administrative context of China's environmental management revealed regulatory and implementing interests fused together to a degree unlike anything found in the other case studies in this book. In the Finnish case study, for example, regulation and implementation of hazardous waste management were *de facto* mixed, but the formal environmental institutions stipulated clear separation of the two. In China, both informal and formal institutions assume that environmental policy, regulation and implementation form a unified and inseparable entity. Chinese concepts and cultural traits provide informal institutional support for the administrative merging of regulation and implementation. The formal environmental institutions in China support a power hierarchy that makes no distinction between regulation and implementation. The hierarchy is strengthened by a regulatory system that rewards the good environmental performers and excludes the poor ones. Environmental management in the Western sense of the term is an absent 'foreign ghost' in China today.

Yet there are opportunities for instituting environmental management practices in Chinese enterprises. Establishing corporate environmental management will become increasingly important in the near future, not just because of the inadequacies of existing environmental protection, but also because China's economy is opening up to Western business. If the differences are not addressed, confusion and frustration are likely, as cooperating partners find themselves using the same language but with fundamentally different meanings.

The tight organizational coupling of environmental regulation and implementation has essentially squeezed out environmental management in Chinese enterprises. If the logic of the preceding case studies were followed, the recommendation here would be administratively to uncouple regulation from implementation. Yet in this case, such decoupling might do more harm than good. After all, one of the most important findings of the contextual analysis was that the fusion of conflicting interests in China is grounded in both formal and informal institutions. Formally to separate regulation and implementation might therefore violate the informal assumptions of society.

There are, however, other ways of satisfying conflicting social interests. When organizational mixing of conflicting interests is necessary for good reasons, such as ensuring the compatibility of informal with formal institutions, contextual balancing of interests is an alternative. Contextual balancing aims to create a forum of discussion, deliberation and decision making over polarized environmental issues. Some authors have proposed that the environmental impact assessment process be developed into a forum of contextual balancing (Taylor 1984). It is interesting to note that the institutional design implications here have much in common with those of the Finnish case study (see Chapter 6), where the public credibility of waste management may require that a public policy making agency also has powers to implement environmental management. These issues will be discussed in further detail in Chapters 8 and 9.

The Sino-Dutch cooperation project successfully kicked off environmental management training at HEMC. In the spring of 1995, I arranged an intensive one-week 'Practical Seminar in Environmental Management' at HEMC, which introduced government officials and company managers in Hunan Province to the basic concepts and latest developments in corporate environmental management, and featured as guest speakers a Dutch sinologist specializing in environmental pollution in China and a representative of the Hunan Province Environmental Protection Industry Association. Since the leadership at HEMC was committed to environmental management training, I also developed in collaboration with LM a fifty-eight contact-hour environmental management course that was included in HEMC's two-year Enterprise Management Programme. After a brief environmental management training period in the Netherlands, LM ran HEMC's first pilot course in environmental management in 1996.

The case study on environmental management in China's Hunan Province concludes the empirical part of this book. The case descriptions have focused on the details of the analytical exercise, but have only alluded to the institutional design recommendations that emerge from the analysis. The next chapter provides a summary of the design principles obtained from the four case studies, and develops them into more specific design guidelines.

Part III

INSTITUTIONAL REFORM PRINCIPLES

8

PRINCIPLES OF INSTITUTIONAL REFORM

The Brundtland Commission's definition of sustainable development (one that meets the needs of present generations without compromising the ability of future generations to meet theirs) has in many ways been adopted as the global principle of sustainable development. The definition, however, has not helped environmental policy makers and practitioners to translate sustainable development into operational environmental management guidelines. On the contrary, many analysts have come to the conclusion that sustainable development is impossible to make operational in any stable way due to its inherently political and dynamic nature (Haila and Levins 1992; Norgaard 1994; Redclift 1992; World Commission on Environment and Development 1990).

Not surprisingly, the decision makers and experts I interviewed in the US, Finland and China from the late 1980s to the mid-1990s were a frustrated lot. In each of the four case studies, the institutional rules of environmental management imposed fundamentally conflicting policy imperatives upon individual policy makers. To make matters worse, the policy makers lacked the autonomy to settle the conflict in accordance with their honest beliefs. Officials, managers, and experts interviewed in the case studies all expressed a clear preference for practising ecologically sustainable management. However, under existing institutional boundary conditions, they felt compelled to adhere to localized, short-term and profit-oriented operating assumptions in policy implementation. Policy makers and managers 'resolved' the resulting dilemma with the cognitively dissonant position of preaching long-term sustainability while practising short-term economics. This state of affairs persists, because the formal institutional boundary conditions and the cognition of individual experts reinforce each other. The more the experts believe that formal institutions guide them to act on a short-run basis, the stronger their cognitive dissonance. But the stronger the dissonance, the more the experts try to weaken it by following, and thus supporting, the existing institutional guidelines.

The policy outcomes of the dilemma are slightly different in the four cases. When the environmental issue at hand is perceived to be acute, as in the Californian drainage case study, policy makers opt for research, development and other such activities that give the general public the impression experts are doing something about the problem. At the same time, they avoid action-forcing, long-term decisions that might raise potentially devastating questions about the rationale of continuing an economic activity that causes serious environmental problems in the first place. When the environmental issue is not yet as acute, as in the Colorado, Finnish and Chinese case studies, policy makers turn completely away from the issue of long-term environmental management in their decisions. Instead, they focus on refining technologies and administrative procedures that ensure compliance with environmental regulations in the short run, but also guarantee the persistence of a wasteful industrial infrastructure. But after all the manoeuvring, sustainable development still looks like an unattainable vision.

The current infeasibility of the concept of sustainable development does not, however, invalidate the objective of searching for institutional configurations leading to sustainable development. After all, many other equally popular and 'infeasible' concepts, such as democracy and justice, have proved powerful shapers of modern societies. It is precisely the socially grounded nature of sustainable development that makes the institutional design so important. From the institutional perspective, the practical challenge for a given society at a given time is not to find a scientifically objective definition of sustainable development, but to develop institutions that enable the society to reach a legitimate consensus on what long-run sustainable development means and how it can be approached.

To overcome the myopic lock-in between institutions and mental models, this chapter will argue that institutional rules need to be rewritten in a way that awards environmental managers and policy makers the autonomy to act upon what they believe to be sustainable environmental management. Now, if the mission of this autonomous group is sustainability, which so effectively escapes objective scientific definition, then one may legitimately wonder why environmental managers and policy makers – i.e. the experts – should be included in the group in the first place. Does expertise have any significance when dealing with an inherently complex and uncertain issue? I think it does.

First, given the complex and uncertain nature of sustainability and its close analogy with democracy, I think the definition of an 'expert' needs to be broadened. There are signs of this already happening. Witness developments in the Arctic, a region which only recently has attracted considerable environmental policy interest. The Arctic regions are home to numerous indigenous peoples with profound traditional and locally sensitive knowledge about human environmental interaction in the region. Recent guidelines on environmental impact assessment in the Arctic, published by the eight Arctic nations, reflect this reality by emphasizing the importance of traditional environmental knowledge in the assessment process (Arctic

Environmental Protection Strategy 1997). For the purposes of a modern environmental management system, local and indigenous peoples have effectively been redefined as experts. Second, having this broad group of experts jointly thinking about sustainability makes sense in the same way as it makes pedagogical sense to have those who are preoccupied by the unknown try to understand it. But, when reading on, one should keep in mind that learning is made of repetitive trials and errors.

Before working out the specific institutional reform ideas, it is useful to summarize the institutional problems observed in the case studies.

THE INSTITUTIONAL PROBLEM

The preceding case studies illustrate a broad range of decision making problems covering all levels of environmental institution. To refresh the reader of the Chapter 2 discussion, institutions of environmental management can be categorized into three levels: operational rules, collective choice rules and constitutional choice rules (Ostrom 1994). Operational rules determine when, where and how to utilize the environmental resource; who monitors the actions of others and how; what information must be exchanged or withheld; and what rewards or sanctions apply to resource use. Collective choice rules have an indirect impact on operational choices. Managers and officials use these rules to form environmental policy and to design environmental management systems. Constitutional choice rules influence the formulation of environmental policy and management agendas, and determine who is eligible to participate in resource use. These rules guide the design of collective choice rules which in turn affect the set of operational rules (see also Table 2.2).

Institutional rules then form a nested hierarchy, in which changes at one level of rules take place under the influence of another set of rules at a deeper level. The nested organization of rules is evident in the mental models the interviews revealed. In each case study, concrete operational and collective choice rules could be identified that had to do with day-to-day policy, management and regulation. But the operational and collective choice rules were always found to be grounded in a deeper set of constitutional rules dealing with the relationship between economic activity and environmental sustainability.

At the operational level, two types of institutional problem were identified. The first is the poor reliability of environmental management systems. Reliability problems caused by poor monitoring and evaluation were particularly serious in the Californian and Colorado case studies. In California, toxic drainwater caught the irrigation bureaucracy by surprise, and its emergency response to the bird deaths and deformities was slow and inadequate. Poor reliability seriously eroded public confidence in the ability of the irrigation bureaucracy competently to handle the drainage problem, because the toxics emergency at Kesterson came in the middle of an already elevated public criticism of the bureaucracy's improper handling of migrant workers, agricultural subsidies and scarce water. In Colorado, even the return flow experts themselves had fundamental gaps in their knowledge of the region's geohydrology, which only fuelled speculation and conjecture in the water management debate.

The second operational level problem found in the case studies is the administrative merging of the regulators and implementors of environmental management. The regulators of Finnish hazardous waste management, for example, are responsible both for the short-term economic viability of the nation's hazardous waste management monopoly and for the long-term environmental sustainability of hazardous waste management policy. When the conflicting duties are administratively combined, the regulators must operate under considerable cognitive dissonance, when fulfilling one duty violates the other. An even stronger version of such environmental corporatism was found in China, where environmental regulators are part and parcel of the organization responsible for implementing environmental engineering in state-owned enterprises. As a result, the regulators think that environmental regulations should be strictly enforced, but also recognize that such action is infeasible because it might entail mass unemployment and endanger China's rapid economic growth.

The problematic collective choice rule found in the case studies was the administrative fusion of the control over different stages of environmental management technology. In the western US, for example, the management of agricultural drainage belongs to the bureaucracy that is also responsible for the management of irrigation. The Colorado and Californian case studies show that, in practice, this puts the administrative emphasis on securing the short-term profitability of agriculture, at the cost of long-term and environmentally sound drainage management. An individual official experiences a cognitive conflict between the necessity to ensure profits for irrigated agriculture and the conviction that the striving for profits also undermines environmentally sustainable drainage management. The result is paralysed decision making, which shields the bureaucracy against political conflict over drainage but also commits it to manage a gradually degenerating agricultural system. Similar dynamics were observed in Finland's Helsinki region, where the semi-governmental waste management authority has comprehensive duties to reduce, recycle, transport and dispose of municipal waste. The organizational fusion of engineering stages with fundamentally conflicting policy implications prevented decision makers from agreeing on any policy at all. The most concrete decision they could agree on was to postpone a final decision on long-term waste management policy to some unknown future date.

Finally, all case studies contain evidence of an underlying problem at the constitutional level, which can be summarized as the institutionalized fusion of the social interest to reorient society toward an environmentally sustainable path of development with the social interest to guarantee the short-term viability and profitability of the society's economic actors. The constitutional design problem becomes evident by virtue of the nested structure of the different levels of environmental institution. Underlying the operational problem of merging environmental regulation with implementation, for example, is a fundamental constitutional problem. It is the assumption that environmental regulation and the implementation of environmental technology could both be made operational as parameters and variables in a single model capable of generating a socially optimal solution. The collective choice problem of fusing together the administration of different stages of environmental technology has its roots in a similar fallacy. It is the assumption that, since waste reduction and waste incineration, for example, are two links in a contiguous chain of engineering operations, they should be managed by a single agency. Yet the case studies in this book, and modern environmental conflicts elsewhere, indicate that neither of these assumptions is correct. Environmental regulation ultimately aims to protect the long-term environmental integrity of ecosystems, whereas the implementation of environmental technology is primarily concerned with the creation of an effective, efficient and profitable engineering operation. And where the reduction of waste implies a profound restructuring and dematerialization of the economy over several decades, the main objective of a waste incinerator is to generate enough energy to keep the operation profitable in the short run. The roots of the cognitive dissonance and circular argumentation encountered in the interviews thus originate at the deepest, most informal level of institution, where the constitutive conflict is between the perceived need for policies that meet short-term economic demands and the conviction that those same policies endanger the ecological foundation of economic activity in the long run.

INSTITUTIONAL REFORMS

The idea underlying all of the following reform principles is the 'demand function' for convictions that was presented in Figure 2.1. The higher the price formal institutions put on individual decision makers to reveal their convictions, the lower the likelihood that individuals will reveal them; and conversely, the lower the price formal institutions put on revealed convictions, the higher the likelihood such convictions will be revealed and acted upon. The price here can be any of the sanctions that formal institutions can impose on an individual expert or decision maker, such as diminished professional prestige among peers, reduced authority to take decisions, or even loss of position altogether. Decision makers and experts in the four case studies in this book are all under strong formal institutional pressure to adhere to short-term economic operating assumptions. But they also hold deep professional convictions about the necessity of tackling issues of long-term environmental sustainability. These convictions all sharply contradict the short-term operating assumptions. To deal with the contradictory goals, they have developed elaborate informal rules, all of which are based on the cognitively dissonant thought patterns described in the previous section.

The main proposal of this study is to introduce consonance between the formal environmental institutions and the informal rules of individual decision makers by reducing the formal institutional price individuals have to pay for expressing their convictions concerning long-term environmental management. In practice, this means separating out functions driven by long-term environmental sustainability goals from those driven by short-term economic profitability goals, and awarding a meaningful degree of institutionalized autonomy and authority to the new social agents endowed with the long-term environmental sustainability functions.

The following institutional reform ideas respond to each institutional anomaly identified in the previous section. They are also presented in the order of increasing difficulty in implementation. As Ostrom (1994) points out, operational rules are in general easier to change than collective choice rules, and collective choice rules are easier to change than constitutional choice rules. The reader should moderate any temptation to view the reform ideas as straightforward recipes for action by putting them in the context of data limitations and what is already known about institutional design. The case studies in this book are synchronic snapshots of four environmental issues that were perceived to be problematic at the time of the interviews. Diachronic understanding of the institutional dynamics in each case can be drawn only from the limited contextual description of environmental management debates that supported the argumentation. As such, the case studies can at most offer insights on actions that might trigger further institutional change. This is particularly true in the light of earlier recommendations by other analysts for the design of governance systems for groundwater and other common pool resources. Successful governance systems are characterized by a dynamic, polycentric order that takes advantage of local specialization and scale and enhances innovation, adaptation and learning (Blomquist 1992; Ostrom 1994). I will return to these issues in Chapter 10 when discussing the implications of the results for future research on environmental institutions.

Increasing the reliability of environmental management

Failures in environmental controls brought irrigation agencies in the western US under embarrassing public scrutiny, which inspired many critics to question the very legitimacy of industrial agriculture in the region. One of the most important first tasks of the drainage agencies (they would, as will soon become clear, be different from those responsible for irrigation) would be to rebuild the seriously eroded public confidence in the ability of any agency to handle agricultural drainage responsibly. This means not only that the drainage agencies – and, for that matter, any agency dealing with contentious environmental management issues – will have to learn how to cope with uncertainties, but also that they must convince stakeholders that the uncertainties can be coped with.

A necessary, though not sufficient, condition for such trustworthiness is that the bureaucracy establish a fairly reliable monitoring and evaluation (M&E) system of pollution, treatment and disposal (Grismer *et al.* 1988; Uphoff *et al.* 1988). Such an agency should be capable of delivering its 'service' as reliably, and as credibly, as those few organizations in our society (e.g. air traffic control and some private utilities) that do manage to carry out their tasks successfully with both a high level of performance and a great deal of public acceptance and credibility (Rochlin *et al.* 1987). The characteristics of existing 'high reliability' organizations include: (1) nearly complete causal knowledge of how technical and management procedures achieve outcomes in the organization; (2) nearly error-free performance from both the organization's personnel and its technology; (3) the organizational capacity to detect deviations from accepted performance standards; and (4) the organization's knowledge of what constitutes a failure to achieve its desired outcomes and the consequences that follow (LaPorte 1987; Rochlin *et al.* 1987).

Since the lack of standards and causal knowledge is precisely what was at issue in the two drainage case studies, high reliability is likely to be attained only over the long run. None the less, making high reliability an explicit goal suggests that the environmental management organization can build up its trustworthiness by being able quickly to detect, and, where possible, respond to problems related to pollution, treatment and disposal. An advanced cradle-to-grave M&E system will not ensure that the agency can always solve the problems. But it will allow the agency to know better where and when problems occur and can in some cases be prevented from worsening.

The Colorado case study, in which irrigation officials and experts were uncertain about the Arkansas River region's geohydrology, is an example of the potential for increasing the reliability of environmental management. To fill the gap in expert knowledge, an M&E system needs to be established for the river basin. At the time of the study, the USGS had already started a basin-wide water quality monitoring programme (US Geological Survey 1990). The Interactive–Accounting Model (IAM) is an appropriate tool for evaluating the basin-wide data. In addition, detailed water quality monitoring programmes should start in areas where the Soil Conservation Service is planning or already conducting on-farm management projects. However, there is currently little capability to evaluate the significance of local data. Serious consideration should therefore be given to further development and use of a detailed local-scale model (an example of such a model is described in Konikow and Bredehoeft 1974). The detailed model should be modified so that it could be used in the Patterson Hollow and other local project areas, and that its output could serve as input to the basin-wide IAM. Only adequate local and basin-wide water quality data, coupled with the integration of a macroscale model (such as the IAM) and a microscale model (such as the Konikow model), would enable the anticipation of basin-wide water quality impact of local irrigation management.

Administrative and procedural uncoupling of long-term regulatory policy from implementation

The close administrative coupling of implementation and regulation, as observed in the Finnish and Chinese case studies (and to some extent in the Californian case), persuades regulators to compromise sustainability goals for the sake of the profit-maximizing operating assumptions of the corporatist policy machinery. Note that the main issue here is conflict between long-term regulatory policy and short-term implementation of environmental management. Industrial processes today are so complex and large-scale that self-monitoring by the implementing firm or agency is often the only feasible and reliable way of ensuring day-to-day compliance with environmental standards. In the Finnish case study, such an arrangement has secured good environmental performance in the nation's hazardous waste management. But it has also stifled consideration of sustainability strategies. I will first present a straightforward institutional redesign idea to remedy the regulatory problem, and will then refine it to achieve a better fit with the specific institutional and cultural conditions of the Finnish and Chinese societies.

The problem of the regulator with a conflict of interest is not new. Wilson and Rachal (1977), for example, posed themselves the question 'Can the government regulate itself?' and discovered the answer was negative. They suggest that large-scale public enterprise and widespread public regulation may be incompatible. Whenever policy makers wish to broaden the scope of public intervention and regulation, they recommend the privatization of day-to-day management. On the face of it, privatized implementation combined with clear separation of regulation from implementation would appear to be a workable remedy to the observed problems. Armed with both the formal responsibility and the administrative design to pursue long-term environmental goals alone, the regulator would no longer feel that goal compromise was the unavoidable price of maintaining position and authority.

A closer inspection of the Finnish and Chinese case studies, however, indicates that straightforward privatization and administrative uncoupling may have to be modified. Environmental management agencies may need to remain public to maintain their credibility and trustworthiness. If a hazardous waste firm, for example, is privately owned and operated, the public may perceive that the profit motive compromises the pursuit of environmentally sound management (Hukkinen 1990; Hukkinen *et al.* 1988). Furthermore, as the Chinese case illustrates, the notion of a conflict of interest between regulation and implementation is rooted in the Western conception of democratic checks and balances and may, as such, be culturally foreign to non-Western societies.

To encounter the problems of privatization, a so-called science model of regulation has been proposed (Taylor 1984). Instead of concentrating on particular regulatory decisions, the science model focuses on organizational learning and emphasizes the importance of continuously growing knowledge within the organization as it makes a series of decisions over time. Policy choices in environmental management would be made after comprehensive public criticism, much as the scientific community chooses dominant theories after mutual criticism and competition between scientists.

An example of the science model in practice is the environmental impact assessment (EIA) process, which is a complex procedure of environmental analysis first conducted within the organization proposing a specific project, and then subjected to public criticism and comment. Public commenting in existing EIA systems can be criticized as being narrow in both form and substance, the most common practices being hearings or commenting periods for the public. However, public involvement in EIA could be refined into a scientifically enlightened political discourse on environmental policy. Environmental management policies would emerge from a process of contextual balancing requiring explicit presentation of the beliefs, both factual and normative, that underlie the proposed policies.

The science model of regulation would be particularly appropriate to the cases investigated in this book. According to Taylor (1984), the following characteristics of the regulatory situation make the science model more suitable than the conventional approach of administrative standards: (1) regulatory goals are ambiguous, unclear, or little agreed upon, which makes statutory guidance unhelpful; (2) the technology for achieving the regulatory goals is uncertain, which generates even further uncertainty in the goals; (3) the regulatory goals involve influencing the actions of many different organizations, employing many different technologies and situated in a variety of environments, all pursuing their socially approved objectives – a situation of such complexity that setting standards becomes very difficult; and (4) the organizations to be regulated are public organizations, which raises sensitive issues of political authority in a pluralist society with separate institutions sharing power. It is clear that the contextual balancing approach could be applied not just in the Finnish and Chinese cases, but in the Californian and Colorado ones as well.

To consider the practical implications of contextual balancing, let us revisit the case of environmental management in China. Authors on economic restructuring in China often offer privatization as the solution to the inefficient and subsidized state-owned enterprises (SOEs), and present town and village enterprises (TVEs), the success story of China's economic growth, as the model for reforms (Cottrell 1995). What these analysts often overlook, however, is that the TVEs are anything but private. They are rather a diverse group of small and medium scale enterprises operating under a mixture of public, private and collective ownership (Christiansen and Rai 1996) – that is, true corporatist creations at the local level! While privatization is the ideal reform on the basis of economic and regulatory literature, it might violate deeply-held informal rules of Chinese culture. I propose an alternative approach based on contextual balancing.

Regulatory and managerial improvements based on contextual balancing should begin in SOEs, because they are likely to play a central role in China's economy in the near future, and have also been the only firms to improve their environmental performance in the past. However, from the environmental management perspective the SOE should be understood as an entity comprising the National Environmental Protection Agency, provincial and municipal environmental protection bureaux, and the factory itself. Given the hierarchical organization of authority, the focus of attention in instituting environmental management systems should be the regulatory official. He or she, after all, is the closest approximation of a corporate environmental manager in the Western sense.

The main challenge, however, of contextual balancing in China would be to respond to the urgent need for a better linkage between the TVEs and the regulatory machinery. In the case of SOEs, such a link exists already in the Environmental Protection Industry Association of China and the provincial Environmental Protection Industry Associations. The role of the provincial associations is to monitor environmental performance in enterprises, provide information on environmental protection to them, and bring enterprises and regulators together. They get their permit to operate from the provincial regulatory bureau, but are funded by the member enterprises. In this parastatal role they conform with the merging of government and enterprise so characteristic of Chinese society, and perform quite effectively. The Hunan Environmental Protection Industry Association, for example, played a central role in organizing and attracting participants to the practical seminar on environmental management for local enterprises and government officials that was part of the case study. The associations may be one of the few existing avenues for stimulating environmental management in TVEs. As such, they could also serve as the most suitable forum for an experiment in environmental regulation based on contextual balancing.

Administrative uncoupling of different stages of environmental management technology

While in perfect harmony from the technological point of view, the economic and political interests of waste reduction, recycling, recovery and disposal do not match. This applies not just to municipal waste management, but to any environmental management technology: the same conflict of interest appears in irrigated agriculture between irrigation and drainage, or in a firm weighing the benefits of increasing its short-term profits with maximum material throughput versus making a long-term investment in clean production technology with minimum material throughput. As the Californian, Colorado and Finnish case studies indicated, this is a particularly serious problem in public environmental management agencies, which typically are not just technical implementors but policy makers as well. When the agency is facing an environmental crisis demanding urgent action, as was the case in California, full devotion to the long-term environmental issue is perceived to threaten the very existence of the agency. The threat to organizational survival stems from the assumption that engineering stages with long time-scales, such as drainage or waste reduction, are an inseparable physical and organizational component of engineering stages with shorter time-scales, such as irrigation or day-to-day industrial production (Hukkinen 1993a, 1993b, 1995a).

In reality, the physical relationship is inherent, but the organizational one is not. Even though agricultural and industrial production often lead to serious pollution problems, both irrigation agencies and industrial enterprises not only can, but traditionally have, functioned without considering the environmental issues of drainage or waste reduction. Serious consideration should therefore be given to building adequate organizational support for the sustainability principles by administratively separating technologies with short time-scales (such as irrigation, industrial production and waste recycling) from technologies with long time-scales (such as drainage, clean technology and waste reduction). Organizations currently responsible for tasks with significantly different time scales should in the future focus on the one they master. Responsibility for the remaining tasks should be given to suitable existing agencies or to completely new ones.

Creating a new administration devoted solely to issues of long-term environmental sustainability has several advantages. First, it would reduce the decision makers' internal goal conflict in the agency currently responsible for technologies with distinctly different time-scales, such as an irrigation bureaucracy responsible for both irrigation and drainage, or a waste management authority responsible for both waste reduction and commercial energy recovery from waste. An agency with conflicting responsibilities has neither the staffing nor the will to pay full attention to sustainability issues. With short-term responsibilities separated out, officials of the new environmental agency would no longer have to compromise their long-term convictions for fear of losing credibility or authority. Second, the new agency with long-term sustainability responsibilities would bear the burden of political opposition to proposed environmental technologies, as well as responsibility for developing alternative remedies. This would isolate political criticism against a short-term activity such as irrigation or waste incineration from that against toxics and other pollutants with long-term impacts. Third, uncoupling would force the new agency to regard pollution and sustainability as primary problems. Existing agencies with a mix of short- and long-term responsibilities tend to treat the crucial problems of pollution and waste accumulation as 'natural' and unimportant byproducts of economic activity. Finally, a separate agency dedicated solely to sustainability issues would gain credibility by balancing the trade-offs between short-term economic profitability and long-term environmental sustainability openly in the public arena. Keeping the trade-offs in public avoids accusations of conflict of interest or private deals made in the backrooms of bureaucracies and corporations (Hukkinen *et al.* 1990).

Organizational uncoupling would not do away with conflicts over short-term economics and long-term sustainability. The new agency would have to weigh the measures it proposes against a number of other issues, which in the case study examples would include water transfers from rural to urban areas, social problems resulting from taking land out of agricultural production, the burden of environmental regulation on agricultural or industrial operations, and bankruptcies of waste recycling and disposal firms that would follow from policies prioritizing waste reduction and clean technology. Attempts to resolve these issues would also stimulate the necessary political debate on the significance of long-term environmental policies in the society. As a matter of social policy, it is better that the tough choices on sustainable development should result from open political debate rather than from unexpected social disruption. Organizational uncoupling would therefore only transfer conflict from within to between organizations, where several conflict management and resolution mechanisms are already in place. Regulatory agencies and the courts have for years been the prime conflict resolvers in Western environmental issues. Where laws prevent resorting to regulators or courts, as is often the case in conflicts between governmental agencies, negotiations and other arbitration mechanisms would be needed (such as the contextual balancing approaches described above).

Once again, a practical institutional redesign proposal can be made. In the light of the preceding analysis, the present configuration of irrigation and drainage agencies in the western US does not seem justified. First, the US Bureau of Reclamation's (USBR) management of both irrigation water supply and return flow disposal is in conflict with the recommendation for organizational uncoupling. The physical connections between the techniques of irrigation and drainage do not predicate close organizational links between the two, especially since the latter are the very source of today's deficient return flow control and management at all levels of government. Second, the US Fish and Wildlife Service (USFWS) has considerable experience and expertise in biological monitoring of potential drainage problem sites. After all, it was the USFWS that first detected the ecosystem failures at Kesterson reservoir, and has thereafter been intimately involved in the US Department of the Interior's reconnaissance studies of potential problem sites in other western US states. In short, the USFWS, not the USBR, is the federal agency that has come closest to fulfilling the crucial requirement of reliability in return flow management.

The USBR's drainage responsibilities and resources could therefore be given to the USFWS. As a result, the USFWS would expand from an environmental monitoring agency to one of comprehensive environmental management, including drainage treatment and disposal. Similarly, state fish and game agencies should be transformed into principal managers of the state's return flows. The specifics of the division of responsibilities between the USFWS and the corresponding state agencies would naturally require further investigation. The reallocation of drainage responsibility would allow the USBR to concentrate fully on water project management, which even without properly handled drainage has become a full-time task for the agency. Combining drainage monitoring and management within the USFWS would enable it to take preventive action – a considerable improvement on the current situation, in which the agency can do nothing but monitor the path to inevitable ecosystem failure (Hukkinen 1991a).

Uncoupling short-term economics from long-term sustainability and developing discursive environmental impact assessment

The administrative reforms proposed in the preceding sections are responses to policy problems observed at a particular level of institutional rule. At the operational level, for example, the lesson to be learned is that the organizational distribution of regulatory and implementing powers has significant influence on the outcome of environmental management. At the collective choice level, the lesson is that the inappropriate administration of technology can lead to significant political and economic conflicts of interest.

But the proposed administrative reforms also raise the spectre of conflict between technocratic rationality and democratic balancing of interests. What guarantees that the proposed autonomous drainage agency in California, for example, will not develop into the kind of monolithic power base that many analysts – and voters – in the state think that the irrigation bureaucracy already is (Reisner 1987)? What guarantees that the values of the experts will represent the values of the population at large; that the experts have sufficient knowledge of issues beyond their expertise; or that the experts can legitimately resolve the distributional conflicts associated with the various solutions? Representation and elitism are issues already tackled by the early systems theorists when they pondered the possibility of allowing a panel of experts to act as a surrogate for future generations (Churchman 1983).

While the issue of technocracy versus democracy does not invalidate the principle of uncoupling conflicting interests in environmental management, it does indicate that the range of application of the design principle needs to be expanded. The following reform ideas at the constitutive level are much more dynamic in nature than the administrative changes.

The constitutional level problem, which can in fact be traced to all institutional levels, is that sustainability issues are systematically considered as constraints to economic policies with a much shorter time-span, but not as autonomous policies in their own right. The most pressing environmental problems today typically cover geographically large areas and require attention spans over several decades, even centuries. The problem is, as environmental and biodiversity management analysts point out, that policies meeting sustainability criteria for local areas and shorter time-scales often fail to do so for larger areas and longer time-scales (Dovers 1995; Groombridge 1992; Wolf and Allen 1995). So, too, in the Californian case study. The in-valley agricultural management solutions that the latest governmental panel developed were, by the panel's own admission (San Joaquin Valley Drainage Program 1990), sustainable on a regional level over decades but not for the entire valley over centuries. Yet both laymen and politicians have been found to show compassion for the environment and even propose sensible plans to abate environmental pollution. But when actual decisions are taken, short-term economic factors override long-term environmental concerns (Opschoor and van der Straaten 1993; Uusitalo 1991). This tension constituted the officials' cognitive dissonance in the four case studies.

The institutional design challenge arising from the constitutional level problem is: how to guarantee autonomy to global, long-term and primarily environmental imperatives in the face of localized, short-term and primarily economic demands? A workable starting position is to respect one of the fundamental findings of institutional economics and economic history, namely that the immediate instruments of institutional change are individuals maximizing utility at margins with the most profitable short-term alternatives (North 1992). To make institutional change work for sustainable development, individuals should be able to maximize utility but do so within an institutional setting that directs the sum and sequence of individual short-term decisions toward the long-term conservation of the environmental resource base. Government taxes are an institutional instrument allowing individual utility maximization within the wider social context. But to serve sustainable development goals, the emphasis of taxation should move from labour to activities that burden the environment (Constanza *et al.* 1995). And to expand the discussion from the narrow maximization of individual utility to the more comprehensive concept of 'satisficing' individual preferences (March and Simon 1994), we should not forget religion as a model for developing environmental institutions. Most religions offer immediate gratification, such as that brought to the Catholic during absolution or the Hindu from bathing in the River Ganges, but have also survived over millennia and convinced large numbers of believers of eternal life. Bringing up taxes and religion when discussing sustainable development is an attempt neither to secularize nor consecrate the concern for the environment, but rather to remind the reader of the existence of institutions that successfully blend together an individual's short- and long-term interests, in terms not only of measurable utility but of unquantifiable good as well.

Another way of tackling the issue of technocracy versus democracy is to develop political discourse procedures in environmental management (this has already been touched upon in the section on the uncoupling of regulation and implementation; see pp. 155–7). The institution of such procedures is likely to result in surprising configurations of policy agreement. A case in point is the 1995 Conference of the Parties to the UN Climate Convention in Berlin, in which the insurance sector and environmental groups, despite fundamental differences in the mental models they used to rationalize global warming, were able to join forces in lobbying industry to reduce its greenhouse gas emissions (*Economist* 1995). Land use planning and environmental impact assessment are two existing procedures that could be developed into institutionalized forms of interaction between diverse interest groups in environmental management. Both already contain procedures for public participation. But to transform impact assessment, for example, into a democratic decision making forum within the regime of environmental management would require a shift from today's project-based assessment to a permanent assessment body with elected members. This would obviously entail careful consideration of mandates between the new body and other elected bodies. The guiding principle of the new body would none the less remain clear: deciding on issues of sustainable development and long-term environmental management.

Economic instruments such as environmental taxes would require political decisions and would thus nicely move negotiations from the closed corporatist arena to the open parliamentary one. This would expose politicians to the tough choices between short-term economics and long-term sustainability. It would also release regulatory resources for the central task of monitoring and evaluating the effects of regulation on environmental quality. Furthermore, the combination of economic instruments and uncoupled technologies is a potentially powerful tool in clarifying agency mission and accountability. Note that the call for clearly distinguishing technologies with long-term environmental goals from those with short-term profit targets does not contradict the widely-held policy prescription of environmental economics that environmental externalities should be internalized. A municipality, for example, could employ two different units to handle waste reduction and waste disposal, while allocating their environmental costs with economic instruments.

DEFENDING AUTONOMY

The institutional problems and reform ideas are summarized in Table 8.1. The reforms are categorized into operational, collective choice and constitutive, which is also the order of increasing difficulty in implementability. Tackling poor reliability, for example, is possible with engineering and management tools that are already well-developed in both managerial literature and practice. In contrast, little professional or academic guidance exists on the best way to launch environmental tax reform (and even less on how to start a religion). Unfortunately, as Chapter 10 will illustrate, even the reforms that according to Table 8.1 should be at the easy end of implementation, have at the outset met with considerable resistance.

Resistance to the reform ideas of Table 8.1 comes as no surprise. The recommendations run against the accepted wisdom of many recent panels

Table 8.1 Summary of institutional problems and reforms in environmental management

Level of institution	*Institutional problem*	*Institutional reform*
Operational	Unreliable systems of environmental technology	Increase system reliability with monitoring and evaluation
	Administrative fusion of implementation and regulation of environmental management	Uncouple regulation of long-term environmental management from its implementation
Collective choice	Fusion of different stages of technology	Uncouple administration of technologies with short time-scales from those with long time-scales
Constitutive	Fusion of short-term economics and long-term sustainability	Institutionalize long-term sustainability as an autonomous social agent separate from agents of short-term economics, and develop discursive environmental impact assessment

on global environmental policy, such as the Brundtland Commission and the Rio Conference, which advise the institutional integration of environmental and economic concerns (Robinson 1993; World Commission on Environment and Development 1990). These reports argue that, since environmental stresses and economic development are linked to one another, institutions, organizations and policies dealing with these issues should also be integrated. The panels view sectoral fragmentation as the chief barrier to sustainability, removable only by putting sustainable development on the agendas of sectoral authorities. But the case studies in this book show that this solution is precisely the problem!

The central institutional problem is not that environmental and economic policy makers do not speak to each other today – they do, often consulting closely, as the preceding case studies have shown – but that long-term environmental concerns have been structurally excluded from virtually all policy making. Yet the temporal dimension is the backbone of sustainability. Instead of putting sustainability on the agendas of sectoral agencies, sustainability concerns in existing sectoral agencies should, according to this study, be identified and given collectively to a separate body endowed with a sustainability agenda. Such an arrangement would not only give prominence to sustainability and lay the groundwork for elaborating sustainable development policies, but could also stimulate cross-sectoral communication, the lack of which the Brundtland and Rio panels correctly identify as a major hindrance to resolving problems of environmental and economic development.

The uncoupling of long-term environmental and short-term economic interests is by no means extraordinary – it has been not just recommended but undertaken before. The separation of irrigation and drainage is nothing less than a divorce of the administrative system of irrigation from the environmentally tuned normative system that currently restricts drainage management. According to Habermas (1975), such a divorce is one way for an organization to avoid a legitimation crisis in the long run. Similarly, the recommendation to make a clear distinction between implementation and regulation is not new (Wilson and Rachal 1977). The case studies highlight the need to make that distinction specifically between short-term implementation and long-term regulatory policy.

Furthermore, the principle of creating autonomous social agents through institutional design has been proposed before and is already much utilized. Beck (1992) thinks that institutional protection of self-criticism in society's specialized fields of expertise (such as environmental management) – organized as alternative evaluations, alternative professional practice, and discussions within organizations and professions of the consequences of their own development – is the only way in which high-risk mistakes with serious unwanted consequences can be detected in advance. In many societies today, institutional autonomy enables judges, professors and voters, for example, to act independently without fearing social sanctions (North 1992; Scott 1987). The creation of autonomous social agents with the mandate to make long-term environmental policy would therefore be an expansion of an already well-established principle in social organization.

One could, of course, resist organizational uncoupling on the basis that what we may be witnessing in the four case studies is a classic situation involving the power of special interests in political systems. Although the exact forms of efforts to capture regulatory agencies or circumvent regulations vary from one political system to another, the general phenomenon of interest group politics is universal. In other words, avoiding the costs of long-term solutions to the drainage problem in California may be fine from the point of view of powerful agribusiness interests, as may be avoiding the costs of air pollution control for commercial entrepreneurs in China. If this were the case, separating functions in organizational terms would simply trigger new opportunities for bureaucratic infighting between agencies that have somewhat different responsibilities in the same broad issue area. And organizations bargaining for short-term profit would surely outnumber the environmentally oriented ones, even after the separation of conflicting interests.

However, such criticism misses the essence of the dilemma that the agencies in the case studies are facing. Avoiding the costs of long-term solutions is certainly not acceptable for the agencies concerned, because doing so undermines the rationale for their existence. By definition, a dilemma requires a choice between two equally balanced alternatives and therefore defies a satisfactory solution. The agencies opt for the short term not as the 'better' alternative, but as the one which formal institutions resist the least. Furthermore, after organizational separation of functions, the game between conflicting special interests would not be one between equals. The proposed regulatory recommendations, for example, increase the autonomy of public organizations that have the power either to determine the limits of legitimate actions in the society, as the parliament and other policy making bodies do, or to make sure that the players stay within the limits, as the regulators do.

A larger concern is that the public sustainability organizations, despite their autonomy, would not gain the authority needed to operate effectively in environmental policy. As the preceding analyses show, short-term economic imperatives are prioritized in decision making because they are deeply embedded in the institutional constellations of US, Finnish and Chinese societies. The embeddedness of short-term rules at all levels of institution in society means that, to a very significant degree, short-term economic power structures constitute these societies. It also means that long-term sustainability interests, if taken seriously, are not just a violation of the short-term economic interests but threaten the entire social power structure that the short-term interests weave together. Put in this perspective, the reluctance of decision makers and experts seriously to consider and put in place long-term environmental policies reflects an understandable reluctance to question the legitimacy of the existing power structure. But it also makes it easy to understand how the new autonomous sustainability organizations might well turn into popular yet marginalized discussion clubs in corporations and in society at large.

There are at least two ways to avoid such marginalization. The agents of sustainability could either play the game with the short-term interests by converting the sustainability concern into a currency whose value the short-term interests accept. Or they could refuse to play the game altogether by adopting a currency that the short-term interests are ill-equipped to cope with. Economic instruments in environmental management are an example of the first approach. As described earlier, once deliberated and decided upon, they exert influence over the long-term direction of social development through incremental, short-term optimizing decisions by individuals. Publicity is an example of the second type of currency. While decision making and expert elites may perceive sustainability concerns as a threat to the existing power structure, the past three decades of public support for the environmental movement in Western industrialized countries indicate that the citizenry at large perceives the threats differently. Publicity may be the reference value the autonomous sustainability organization needs to rely on to gain and maintain authority. I will return to this issue in Chapter 10.

Yet another criticism against the recommendations of this book is that the public choice processes that were observed may constitute accurate reflections of individual and societal values. In other words, the preference for long-term sustainability, as expressed by the public and policy makers, may not be revealed in the economic sense of the word. Once the costs or trade-offs of actions taken for the long term become clear, there might (or might not) be less support for sustainability. This is a generic problem for public non-material goods, such as environmental quality, because there is no market in which to reveal the preferences. In fact, some methods available for obtaining such preferences, such as contingent valuation, must rely on the assumption of a hypothetical market: respondents are asked how much they would be willing to pay if a market existed for the environmental good in question (Pearce and Turner 1990).

But in the case studies in this book, institutional factors are precisely the veil that hides preferences for the long term. Others have arrived at the same conclusion. Uusitalo (1991), for example, has shown in her study of Finnish consumer attitudes toward the environment that, while Finns value greatly environmental quality and protection, the materialization of this preference as action will require collective agreement on norms of behaviour to guarantee that an individual's investment in environmental quality will not be used up by free riders. In other words, institutional reform of the type proposed here is the prerequisite for revealing the sustainability preference.

Now, assuming sustainability preferences have been revealed as a result of novel institutions, I have no illusions about the experts being able to resolve the issues of short versus long run in terms of a commonly perceived reality that would allow for a definitive benefit–cost analysis. In complex environmental debates, all sides can be perfectly well structured and informed, yet each can arrive at a very different calculus of the costs and benefits. The point is that the proposed institutional reforms aim to encourage well-developed and openly argued alternatives for both the short and the long term, which would put us in a better position than we are today to resolve disagreements between them. As the case studies show, experts today tend to eliminate long-term alternatives already at the cognitive level, giving sustainability concerns little chance of ever being elaborated as concrete programmes of action at a level of detail comparable to their competing short-term alternatives.

One of the key tasks for future research is to look for historical cases of institutionalized advocacy for the long-term interest of an environmental resource. Such cases could serve as prototypes or indicators of what institutionalized sustainability might mean in practical terms. Today's environmental management contains the seeds of such institutions, as the next chapter will argue on the basis of examples from public agencies and private corporations. To assist future designers of environmental institutions, the design recommendations of this study, which relate to contemporary environmental management systems, will be compared with those formulated for more traditional natural resource management systems.

9

INSTITUTIONS OF INDUSTRIAL ECOLOGY

Institutional change is path dependent. It is impossible to understand institutions without understanding the historical process by which they were produced (Berger and Luckmann 1967). What is more, the historical process by which we arrive at today's institutions constrains our ability to make them fit for the future (North 1992). This has significant implications for the design of institutions of long-term and large-scale environmental management. The design principles we draw from an analysis of long-surviving environmental management institutions are constrained by the initial conditions and development processes that have moulded the institutions into what they are today. The principles may also be incompatible with the ones we arrive at by analysing the institutions of modern environmental management, which have much shorter life histories and fundamentally different initial conditions.

This chapter compares institutional design principles derived from analyses of long-surviving, robust environmental management systems with those formulated on the basis of the relatively short-lived, modern environmental management systems treated in this book. Drawing from extensive empirical case evidence on long-surviving, self-governing institutions, Ostrom (1994, 1995) proposes design principles for institutions of long-enduring common pool resource management systems. Ostrom examines mountain grazing and forest management systems in Switzerland and Japan and irrigation systems in Spain and the Philippines. The youngest of the long-surviving natural resource management systems Ostrom analyses is more than 100 years old, while the oldest one exceeds 1,000 years. Since the design principles are based on empirical evidence of institutions that date back centuries, they bring together information about those characteristics in the initial conditions and evolution of old institutions that have permitted them to adapt to changing social conditions throughout history. But they tell us little about the initial conditions and changes required of institutions that have emerged in the modern industrial age and that we would like to survive far into the future.

The design principles of this book are derived from case studies of what I have termed modern environmental management. First, the time-scales are of a different order of magnitude from those in Ostrom's study. The Chinese environmental management systems studied in this book have begun to emerge over just the past decade or so, and the centralized irrigation bureaucracies in the western US have existed for approximately half a century, since the construction of the big water projects.

Second, modern environmental management takes place in competitive and often global markets for goods and services. Short-term profit maximization is the rule of operations, increasingly also for organizations such as public utilities that used primarily to provide a public service. The natural resources these operations use are frequently public goods. It is well established that markets do not perform effectively in the allocation of public goods where exclusion is difficult, as is often the case with natural resources (Ostrom 1995). However, the global market is such a strong boundary condition – witness its central role in major social and economic restructuring programmes around the world (Reed 1996) – that it would be unwise to disregard it when considering environmentally oriented institutional change. Furthermore, the resources are not always spatially localized. The California water supply system, for example, draws water through channels and pipelines from the Colorado River, which in turn has implications for water management in virtually every arid state in the western US. The stakeholders of modern environmental management are even more dispersed. In California, they include not only the water users of all the water systems that the California system has linkages with, but also a number of environmental interest groups, irrigators, water systems planners and environmental regulators throughout the US (see Chapter 5).

Finally, and in contrast to Ostrom's analysis of success stories of robust institutions, my case studies focus on failures of institutional design: why have modern environmental management institutions failed? Why do the policy makers involved think they have failed? And how do they think the situation should be corrected? Thus the design principles are not based on empirical evidence of successful institutions of modern environmental management, simply because we cannot yet say if such exist. Instead, the empirical analysis focuses on the central process of institutional change, namely the feedback between the mental models policy makers use to understand issues and the institutional set-up within which they try to resolve those issues.

The case studies in this book are about modern environmental management systems. The market, the drive toward profit maximization, and poorly defined management system boundaries are significant boundary conditions for irrigation managers in California and Colorado, waste managers in Finland, and increasingly also for environmental managers in China. At the same time, these managers are under considerable social and political pressure to adhere to the principle of sustainable development, however it may be understood. These pressures come from a combination of publicly acknowledged failures of past environmental management and fear for future management failures if current policies do not change.

The design principles to emerge from the case studies are quite different from those developed by Ostrom. However, when considered in the light of recent developments in environmental management, particularly the prescription to develop so-called industrial ecology as a practical step toward sustainable development, the two sets of design principles complement each other. I will first summarize the design principles developed for long-enduring institutions of natural resource management and then discuss how they fit with the design principles of modern environmental management in the context of industrial ecology.

INSTITUTIONS OF LONG-ENDURING NATURAL RESOURCE MANAGEMENT

Ostrom (1994) identifies eight design principles that characterize long-lasting, robust natural resource management institutions. The first two deal with the boundaries of the resource and its users. First, individuals or households who have rights to withdraw from the common-pool resource must be clearly defined, as must be the boundaries of the common-pool resource itself. If the boundaries are not clearly defined and the outsiders effectively closed out, local appropriators face the risk that the benefits they produce by their efforts will be reaped by others who have not contributed to those efforts. Second, appropriation rules restricting time, place, technology, and/or quantity of resource units are related to local conditions and to provision rules requiring labour, materials, and/or money. Appropriation rules concern the allocation of a fixed, time-independent quantity of resource units, whereas provision rules concern the assignment of responsibility for building or maintaining the resource system over time. Rules tailored to local conditions have been found to account for the perseverance of the resource management system.

The next three rules deal with the regulatory setting. The third rule states that most individuals affected by the operational rules can participate in modifying the operational rules. Fourth, monitors, who actively audit common-pool resource conditions and appropriator behaviour, are accountable to the appropriators or are the appropriators. And fifth, appropriators who violate operational rules are likely to be assessed for graduated sanctions by other appropriators, by officials accountable to these appropriators, or by both. In short, the design and monitoring of environmental regulations and the sanctioning for non-compliance with those regulations should, to a significant extent, be the responsibility of those regulated.

The next two rules deal with self-determination of appropriators in conflict resolution and organization. According to the sixth rule, appropriators and their officials have rapid access to low-cost local arenas to resolve conflicts among appropriators or between appropriators and officials. In other words, for individuals to follow rules over extended periods of time, there must be some mechanism for discussing and resolving what constitutes an infraction. Seventh, the rights of appropriators to devise their own institutions are not challenged by external governmental authorities. Resource users should thus have the possibility of devising their own rules without creating formal governmental jurisdictions for the purpose.

Finally, the eighth rule states that appropriation, provision, monitoring, enforcement, conflict resolution and governance activities are organized in multiple layers of nested enterprise. Resource users are typically organized as nested levels, as are the institutional rules. Establishing rules at one level without considering the rules of the other levels will produce an incomplete system that may not endure over the long run.

As the next section will show, there are marked differences between these principles and those crafted for modern environmental institutions.

TOWARD INSTITUTIONS OF INDUSTRIAL ECOLOGY

Table 9.1 is a comparison of Ostrom's (1994) institutional design principles with those developed in this book. To refresh the reader, the design principles outlined in the previous chapter had to do with increasing the reliability of environmental management systems, uncoupling long-term regulation from short-term implementation, uncoupling the administration of short- and long-term technologies, developing discursive environmental impact assessment and institutionalizing long-term sustainability as an autonomous agency. The two sets of principles deal with the same broad issues of institutional design, namely the boundaries of resources and stakeholders, the regulatory setting, conflict resolution and the overall context of governance.

While there are differences in the details of the two sets of design principles, the differences, when viewed in light of recent developments, such as ecological restructuring of industrial production, present themselves as nuances of broadly coherent recommendations. I will explore these developments and their institutional implications in four parts. First, I will outline the ecological restructuring project in the business world in the industrial ecology framework. I will then detail three areas of product strategy in which industrial ecology has already had a dramatic influence and will do even more so in the future. The central element of an environmental product strategy is the development of environmentally sound product concepts, which implies a distinction between the material product

Table 9.1 Comparison of institutional design principles derived from long-surviving common-pool resource management institutions with those derived from modern environmental management institutions

Focus of design	*Design principles derived from long-surviving, robust institutions of resource management*[a]	*Design principles derived from modern environmental management institutions*
Boundaries of resources and stakeholders	(R1) Individuals or households with rights to withdraw resource units from the common-pool resource and the boundaries of the resource itself are clearly defined. (R2) Appropriation rules restricting time, place, technology, and/or quantity of resource units are related to local conditions and to provision rules requiring labour, materials and/or money.	(M1) The reliability of environmental management is increased by establishing permanent monitoring and evaluation systems. (M2) Those stages of environmental management technology that are primarily driven by economic profit objectives are identified and administratively separated from those stages guided by long-term environmental management objectives.
Regulatory setting	(R3) Most individuals affected by operational rules can participate in modifying operational rules. (R4) Monitors, who actively audit resource conditions and appropriator behaviour, are accountable to the appropriators and/or are the appropriators themselves. (R5) Appropriators who violate operational rules are likely to receive graduated sanctions (depending on the seriousness and context of the offence) from other appropriators, from officials accountable to these appropriators, or from both.	(M3) Long-term environmental policy is the responsibility of an autonomous agency which is institutionally separated both from the implementors of environmental management and from the environmental regulators concerned with short-term regulatory compliance.
Conflict resolution	(R6) Appropriators and their officials have rapid access to low-cost, local arenas to resolve conflicts among appropriators or between appropriators and officials. (R7) The rights of appropriators to devise their own institutions are not challenged by external government authorities.	(M4) Environmental impact assessment as a scientifically enlightened political discourse on environmental policy should guide the creation of political procedures for settling environmental disputes, particularly between public agencies with conflicting missions.
Overall context of governance	(R8) Appropriation, provision, monitoring, enforcement, conflict resolution and governance activities are organized in multiple layers of nested enterprises.	(M5) Those elements of the environmental management system that aim at long-term environmental management are identified and organizationally separated from those elements concentrating on short-term economic profitability.

[a] *Source*: Ostrom 1995

and the service it provides. The aim of such abstraction is to reinvent a product in a way that provides approximately the same service with a radically diminished burden on natural resources and the environment. Throughout the discussion I will point out ways in which firms striving for industrial ecology are likely to find the two sets of design principles complementary to each other.

Industrial ecology

The challenge ecology poses to industrialism is to transform the relationship between humanity and nature from exploitation to imitation. This new relationship has been called industrial ecology (Tibbs 1992), industrial ecosystem (Frosch and Gallopoulos 1989), or industrial metabolism (Ayres 1989). The imitation can have a wide spectrum of modalities. At one end, industrial production can replicate the energy and material conversion processes found in ecosystems. An example is the waste water treatment process that removes nutrients by growing algae in waste water with solar energy (Oswald 1991). Full-scale applications already exist in many sunny regions of the world. At the other end, industrial processes must be designed to mimic only the careful accounting of matter and energy found in ecological systems, while operating in complete isolation from them due to systemic incompatibilities. Our nuclear inheritance, for example, gives us no option but to develop such isolated systems (Krauskopf 1990).

Industrial ecology has common elements with such ecological analogies as 'organizational ecology' and 'business ecosystem' used in organization and business literature (Hannan and Freeman 1977; Moore 1993). All of these concepts view companies as parts of larger networks of organizations with cooperative and competitive interactions. But industrial ecology is different from the rest in a very important sense: it focuses on the exchange of physical matter, energy and information between the members of the industrial ecosystem, something the others disregard. Much recent literature has focused on the physical complexities of material and energy flows in industrial ecology (Allenby and Richards 1994; Ausubel and Sladovich 1989; Ayres and Kneese 1989; Côté *et al.* 1994). The institutional and organizational complexities involved in the management of an industrial ecosystem have received much less serious academic attention.

But since we are concerned about the reliability and ultimately the survival of complex technologies such as industrial ecosystems, our primary attention should be on the human systems designed to manage them (Morgan 1986; Perrow 1984; Vickers 1983). We can afford to observe with casual indifference the rise and fall – the latter frequently resulting from inappropriate institutions or management – of experiments to recycle such benign items as plastic bottles or newspapers. The luxury of trial and error is lost, when the social decision is taken to establish comprehensive systems to recycle hazardous or nuclear wastes. To operate a highly complex technology with tightly coupled organizational linkages (such as a recycling system), and do it reliably without accidents, puts an extraordinarily heavy burden on the design of appropriate human systems (Perrow 1984; Rochlin 1990; Rochlin *et al.* 1987).

The complexity of the systems and interactions involved in industrial ecology is staggering, as the following analytical perspectives illustrate. Like ecological production, industrial production has a metabolism, which refers to the rate of conversion of material and energy within, say, an industrial plant. This is one framework for analysing the industry. But it can also be analysed in terms of industrial ecosystems, which focus on the exchange of energy, material and information between the industrial players. Then the focus can shift to the interface between industrial production and natural ecosystems. What are the possibilities for reducing natural resource extraction through material and energy conservation in an industry? And what are the possibilities for shifting to a low- or non-waste technology (Tibbs 1992)? The inherent systemic complexities of industrial ecology are compounded by the analytical complexities involved in conducting a life-cycle analysis (LCA), which aims to report the cumulative environmental impact of a product throughout its life-cycle. Thus we are still waiting for the final scientific word on which are sounder for the environment: cotton or paper nappies; PVC, polystyrene or polypropylene containers; recycled aluminium cans or return glass bottles; and many more. Finally, the complexities of industrial ecology and the consequent analytical confusion can have a paralysing effect on decision making, when social groups with diverse political and economic agendas use conflicting mental models to understand the system (Fairclough *et al.* 1993). Scientific uncertainties and complexities frequently open up the platform of both public and corporate environmental politics for yet another LCA expert who can question the 'scientific' validity of all previous LCAs.

This is a nightmarish situation for the organizational strategist trying to design more effective boundary-spanning activities to shield the corporation against environmental uncertainties and liabilities. Modern environmental managers should be able to maintain the boundaries of the resource, determine who is eligible to utilize the resource, and relate the rules of usage to local conditions (principles R1 and R2 in Table 9.1). Furthermore, the complexity, uncertainty and indeterminacy of system boundaries increase the importance of ensuring the reliability of modern environmental management systems (principle M1). But how does one span organizational boundaries, when the boundaries themselves are indeterminate, the tools for delineating them give conflicting answers, and the cognitive frameworks within which the tools are used are mutually inconsistent? Note that by poorly defined boundaries of modern environmental management systems I mean *de facto* boundaries, not legally stipulated ones. It is therefore entirely possible that both the legitimate users and the physical boundaries of the resource have been clearly defined, but that the global market context creates avenues for other stakeholders, such as globally operating environmental groups, to boost the success or threaten the survival of the management system, unlike any formal stakeholder.

The temptation is to call for more research to reduce the uncertainty. This, however, only predestines corporate strategists to the very game from which they are trying to escape, namely, that of repeatedly disqualifying all preceding LCAs and environmental impact assessments (EIAs) with the latest 'standard-setting' method.

The alternative is to change the corporation's approach toward ecological issues altogether. Corporations should look at the environmental challenge not as a task of finding and managing corporate boundaries but rather as one of internalizing environmentally sound elements into the company's product. The agency or enterprise should reconceive many aspects of its operation, such as services, regulators and customers, which in the traditional view belong to its task environment, as parts of the organization and its core products. The benefit in the change of perspective is that the diversity and uniqueness of LCA, EIA and other environmental management procedures become the sources of competitive advantage for the corporation.

Service as a product

Although LCAs on the same product may contradict each other, the overall direction of ecological product development is clear. Manufacturing has, at an accelerating pace, transferred matter from nature to artifacts in the economy, and, in doing so, has more often than not struck at the Achilles' heel of nature. Gains in artifact diversity take place at the cost of biological diversity (Pantzar 1992). The objective should therefore be to dematerialize production, that is, to minimize or avoid the material and energy flows of raw material extraction, manufacturing, distribution, use and disposal. Dematerialization and service-orientation are not just ecologically wise strategies, but often sound responses to competitive pressures as well.

The ideal ecological product concept follows as closely as possible the fundamental but often-forgotten notion of economics that the basic economic value is service, not material good. As Ayres and Kneese put it, already over a quarter of a century ago, 'material objects are merely the vehicles which carry some of these services, and they are exchanged because of consumer preferences for the services associated with their use or because they can help to add value in the manufacturing process' (Ayres and Kneese 1969: 284).

Dematerialization is already a reality in many industrialized countries. As a crude measure, the share of services in GDP grew from around 50 per cent in mid-1960s to more than 60 per cent today in such high-income economies as Hong Kong, Australia, Austria, Denmark and Germany (*World Development Report* 1992: 223). Furthermore, industrial economies have grown while the resources and energy used per unit of growth have declined. Chemical companies, for example, have doubled output since 1970 while consuming less than half the energy per unit of production (Schmidheiny 1992: 10). Energy used per dollar of GNP diminished by 31 per cent in Japan and by 23 per cent in the US between 1973 and 1985 (Herman *et al.* 1989: 59).

There are promising examples of product concept evolution from material to service in an industry which brings the modern consumer probably closer than any other to natural resource extraction – the public utility sector. The increasing political and economic cost of building power plants and dams has persuaded some US and European energy and water utilities to 'buy back the resource' (I am indebted to Darrell Ament for the concept), i.e. invest in energy and water saving instead of new capacity (Cairncross 1991; *Economist* 1991; *MWD Focus* 1989). Other public utilities, like waste management agencies, could do the same. Funds reserved for the construction of a new landfill site could purchase consulting services in clean technology for the most wasteful local industries; or buy recycling equipment for individual firms and households; or be invested in informing and educating citizens about waste reduction and recycling. Similar developments have also been observed in corporations (Hukkinen 1995b).

However, the invention of novel combinations of services and the creation of a service industry require considerably more time than the development of conventional production of material products. To speed up the process of dematerialization, public utilities and private firms need to consider administratively separating out those stages of environmental management technology that rely on long-term sustainability objectives, simply to prevent the internalization of conflicting objectives and the problems this study has found to be associated with it (principle M2 in Table 9.1). It should be noted that the goal conflicts between engineering stages become polarized in governmental or quasi-governmental agencies, such as irrigation agencies or regional waste management authorities, which are constantly exposed to the uncertainties of their organizational environment. The agencies confronting these problems are typically not just policy makers with a public service to provide but also technical implementors with economic performance objectives to fulfil. This places a considerable burden on the performance of political conflict resolution mechanisms, such as environmental impact assessment (principle M4). While the private corporation does face the same goal conflict, it can, in the end, resolve it as a multi-objective optimization problem. Finally, the firm or utility should have the authority to define on its own the institutional setting necessary to support the novel service (principle R7). Government authorities, for example, should not stand in the way of innovation by hindering the expansion of a public utility into unconventional areas of service.

Regulation as a product

Recent trends in environmental regulation lend themselves well to reconceiving regulatory measures as part of the product of an enterprise or public utility. Hand in hand with our increased understanding of the complexities involved in human interaction with ecosystems goes an increase in the resources required to regulate that interaction with traditional regulatory approaches. As a result, public environmental regulators are beginning to rely more and more on self-regulatory schemes (Beaumont *et al.* 1993). Both corporations and their regulators need to shift to higher levels of complexity in environmental management systems control. Corporations are beginning to monitor, evaluate and correct their environmental performance themselves – tasks which used to belong to the governmental environmental regulator. They should see these added responsibilities not as a burden but as a new competitive arena, in which the credibility of the LCAs, EIAs, environmental audits and monitoring that went into the production of the product is a central feature able to attract customers.

Once self-regulatory schemes are in place, firms view regulatory agencies increasingly as clearing houses for the latest environmental management innovation. When regulators are valued more for what they can tell about competitors than future regulations, the first step has been taken toward internalizing the regulator into the product concept. Lobbying the regulator to tell you what you would like to hear has become a secondary imperative. The more urgent task is to use the regulator as one of the most reliable scanners of potential ecological product niches in the competitive business environment. The ecological product concept to emerge from such scanning does not necessarily take the form of a tangible improvement in the production process or a radical dematerialization of the product. It can also be convincing, scientifically supported information about the ecological impacts of the product during different stages of the manufacturing chain. Consistent presentation of such LCA information to the consumer in product marketing creates a completely new competitive arena.

Nadaï's (1994) description of the evolution of pesticide regulation in the European Community since the 1970s provides a more detailed model of the driving forces and dynamics of industry participation in regulation. The involvement of one interest group in the regulatory process is perceived as a threat by the competitors, who consequently decide to participate. Anticipation of negative regulatory pay-offs therefore triggers sequential involvement of the subjects of regulation in regulatory deliberations.

In the new regulatory system, environmental regulators assume the responsibility of verifying the reliability of the various self-regulatory schemes. For them, self-regulation is not a matter of losing power over the subjects of regulation, but of exercising that power at a higher systemic level of control and with much longer time-scales than is the case with day-to-day operations of the firm. The differences in the institutional design principles relating to the regulatory setting need to be viewed in this context. When what is at issue is the day-to-day regulatory control, which compares environmental monitoring data with performance standards and corrects the performance accordingly, self-regulation by the subjects of regulation is necessary, because of the sheer complexity of modern environmental management systems (principles R3, R4 and R5 in Table 9.1). But this should in no case mean that environmental regulators become the implementors of environmental management. Long-term environmental policy and regulation, which is a task fundamentally different from day-to-day regulatory compliance, should be the responsibility of an agency endowed with a significant degree of institutional independence (principle M3).

Customer as a product

Waste recycling has proved to be a tenuous business in many industrialized economies. Despite an abundance of waste, sorted good quality waste is simply not around in adequate quantity and acceptable price for well-functioning markets. The only way for a recycling operation to compensate for the exceptional uncertainties in supply and demand is by investing in contracts that secure future transactions. The more attractive, comprehensive and long-lasting the company's contractual packages with its suppliers and customers, the higher its chances of business success. In short, customers are an essential part of the product, worthy of the kind of attention the firm pays to product development.

The arrangement may look like a cartel to the orthodox economist, but it is an insurance against business failure to the technical operations manager. When a company makes plastic film out of bulk polymers, for example, it ordinarily has the luxury of choosing the most affordable of a number of suppliers. When waste plastic is the raw material, the company's survival is tightly coupled with a continuous and reliable stream of good quality waste. Dutch plastics recyclers have created a complex and fragile orchestration of contracts and agreements with their customers. The complex logistics of getting the recyclable waste to the recycling plant adds to the challenge. Bulk polymers usually come from a few central locations, whereas recyclable waste has to be collected from geographically scattered locations. This leads to another point that will incense the orthodox economist even further. The contracts between waste suppliers, recyclers and customers should preferably be geographically localized. Creating complex worldwide logistics for recycled waste is economically and ecologically infeasible. It is usually more economical to rely on local suppliers of recyclable waste than to haul it across the globe. Furthermore, international transport just transforms the localized, and quite manageable, ecological burden into a global, and virtually unmanageable, ecological stress along the transport route in the form of air pollution, congestion and route construction (Hukkinen 1995b).

But recycling is not the only industrial sector where contractual networks are vital. Östlund (1994) reports that similar arrangements emerged within Swedish refrigerator, flexible foam and circuit board industries in response to regulatory pressures to phase out the use of chlorofluorocarbons (CFCs). Collaboration among industries in the production chain was found to be the standard solution to the environmental problems caused by CFCs. Mobilization of resources and coordination of activities to deal with CFCs took place in the context of a network that diffused and legitimized the chosen solutions, both among network members and in the political community.

Networks and markets are not in conflict. The challenge to recycle plastics or phase out CFCs forced the industrial players to act as members of an industrial ecosystem and benefit from the systemic characteristics of geographical localization, cooperation, and long-term commitment. These properties do not in themselves resist the market. In fact, once customer contracts are included in the product concept, the invisible hand can deliver its blessings. Contracts that best fulfil the needs of the customers will survive. To put it in terms of institutional design instructions, environmental managers should have access to local negotiation and conflict resolution arenas in which to design and agree on long-term contracts between the various stakeholders of environmental management (principles R6 and M4 in Table 9.1). Furthermore, the principles of localized control over a localized resource (principles R1 and R2) apply, both to guarantee that the material cycles of the industrial ecosystem do not excessively burden the natural ecosystem and to give the users freedom of innovation in creating the networks. However, the time-scales of the technology with which the resources are going to be utilized still need to be kept in mind. Technologies with short time-scales need to be administered and managed separately from those with longer time-scales (principle M2).

MODERN TRADITIONS?

There are significant differences between the two sets of design principles. In Ostrom's set of principles, the overall context of governance is consistently sensitive to local scale, with large-scale governance structured as nested local systems. This study is less concerned with the spatial scale and emphasizes the identification of those dimensions of an environmental management system that have to do with the long term. The boundaries of the resource pool, its users and the technologies applied in resource extraction should in Ostrom's scheme be clearly defined in accordance with local conditions. The principles presented here dictate that clear boundaries need to be drawn between the users and technologies concerned with long-term environmental management and those dealing with short-term resource use. Ostrom recommends a largely self-regulatory system in which the setting of rules, the monitoring of compliance and the determination of sanctions are the responsibility of local resource managers. This study proposes the creation of an autonomous body with a mandate to establish long-term regulatory policy. The only obvious common ground between the two sets of design principles has to do with conflict resolution, where both systems favour an arena sensitive to the stakeholders and the specific issues concerned (Table 9.1).

I do not consider the differences to be fundamentally incompatible. The concept that unifies the two sets of design principles is autonomy. Both sets prescribe the uncoupling of resource management systems from the more turbulent task environments in which they operate (Roe 1996; Scott 1987; Thompson 1967). The central challenge is to specify accurately the initial institutional setting we are trying to redesign by determining the dimension in which management systems are most vulnerable to external turbulence. Is it a setting in which organizations and individuals are struggling to cope with a pull between short-term profit and long-term sustainability? Or is it a setting in which organizations and individuals have a clear orientation toward a localized community?

The former case would lead us toward the principles described here for modern environmental management, the latter toward the ones Ostrom outlines. Modern environmental management rarely takes place within a self-contained, localized community. Rather, it takes place in the context of a market economy – often a global one – in which neither the environmental resources utilized nor the utilizers and other interest groups are locally specified. While there may be a 'community' of stakeholders, it is often spatially dispersed and bound together by interests other than those of a traditional, spatially localized community. This chapter has identified some of the dimensions where cleavages and clusters in modern environmental management interests are most likely to be found. For these systems, the relevant institutional design principle is temporal autonomy guaranteeing decision making powers over long-term environmental policy. In contrast, spatial autonomy is the central institutional principle holding together a community that has concerned itself with resource management over a very long time. The principles of clearly defined boundaries of resources and their users, self-regulation and graduated sanctions, and localized conflict resolution all rest on the assumption that the people involved in environmental management share common values about how their region should develop and operating assumptions about how it will develop. In these cases, spatial autonomy appears to be the guarantor of long-term survival.

Without sensitivity toward the initial conditions of institutional design, serious incompatibilities can arise. When, for example, the principles of self-regulation and self-monitoring have been applied in Finnish hazardous waste management, the result has been a short-sighted policy by the regulators to guarantee a constant flow of waste to the plant they manage (see Chapter 6). The same principles that guarantee the long-term survival of a local resource management system thus secure, when applied in a different institutional setting, the persistence of an industrial production system that precludes waste reduction and other measures toward long-term ecological restructuring.

A more vexing question is how to create an institutional setting capable of facilitating the emergence of two very different types of environmental management organization for which the preceding analyses show a clear need. On the one hand, the complexities and uncertainties of the sustainability challenge require an organization that constructs itself through participatory dialogue and learns through trial and error. On the other hand, contemporary environmental management systems often operate under a stringent requirement for high reliability, with little or no tolerance for trial and error – simply because the consequences of a single error may be catastrophic, for example a nuclear power plant accident (Perrow 1984). Full and complete causal knowledge of the complex management system is a prerequisite of continued reliable operation, which only a community of highly specialized technical experts can achieve (LaPorte 1987; Rochlin *et al.* 1987). We are back to the issue of technocracy versus democracy, and it appears that future environmental institutions should facilitate both modes of organization. Suffice it to say here that rapid environmental changes with potentially drastic impacts impose stringent reliability requirements on environmental institutions and the management systems they support. Such reliability should be a top research priority, particularly since historical analyses of past climate changes indicate that dramatic environmental changes over a period of decades are a distinct possibility (Taylor *et al.* 1997).

The communities emerging from the design principles outlined here for modern environmental management would be bound together by the long-term interest. The challenge for the future is to integrate the institutions that have persisted in time, and thus have their roots in history, with those that are the products of modern times and strive to persist for centuries to come. It is not just a research challenge, but a practical one as well.

10

EXPERTS IN PUBLIC

My bias in writing this book has been towards the long-term future. It is not just my own personal bias. The decision makers and experts I had conversations with were deeply concerned about long-run sustainable development. They had very different approaches to the future, reflecting, in fact, much of the thinking found in planning and futures research literature (Godet 1993; Mannermaa 1991). Some were planning optimists, who felt that what was at issue was taking control of the future. All that was needed in their mind was a clear vision, translated into strategies, each with detailed action plans. Others took quite the opposite approach, viewing everything in society and nature as an evolutionary process on which human beings can have little influence. On the few occasions on which they might be able to control events, things usually turn out worse than they would have done without human intervention. So, in the end, they felt it was better to let the chips fall where they may. Then there were the systems thinkers, who accepted the notion that social and ecological phenomena are complex, non-linear and evolutionary processes, but who also felt that every now and then opportunities arise in the flow of events for humans to change the direction of development. To them the process is something like driving on a foggy road at a slow enough speed to be able occasionally to see the road signs and then take precautions or choose alternative routes (de Boer 1995).

But I found that the interviewees all had their own road maps which they relied on to understand what the environmental issue was and how it might be dealt with. I also found that these mental models were always mutually inconsistent when considered together, and often internally inconsistent when considered individually. Choosing one or more of the models as the accurate and reliable description of the environmental management problem would have been unfounded, because it would have implied that the rest of the expert models were somehow inferior to the chosen ones. It would also have been unwise, because it would just have polarized the already charged debate. If any intelligence is to be extracted from all this contradictory expert knowledge, it must be the result of an understanding of both the differences of thinking among several experts and the cognitive dissonances of individual experts.

My understanding of the dissonances of expert opinion is based on an analysis of the causalities and contrasts of what the experts told me. I analysed their stories both individually and as groups of individuals. The understanding I have is systemic in two ways. First, in the sense described above, namely that whoever attempts to do something about long-term environmental problems needs to accept social and ecological phenomena as complex non-linear systems. But there is a systemic level encompassing the very act of someone attempting to do something about the problem, namely the level of cognition. The range of options available for policy makers to tackle a complex environmental problem is limited by their shared knowledge of the problem. When the knowledge is shared by a significant number of individuals over time, it becomes an institution, which both reinforces and is reinforced by the knowledge of individuals. The feedback between the mental models of decision makers and the institutions of environmental management is a tricky process. Maintaining the cognitively dissonant mental models in the Californian drainage case, for example, meant that individuals had constructed elaborate and intellectually challenging explanations for the smart ways in which they had learned to play the game.

The subtleties of the feedback between mental models and institutions are both evidence of and a challenge to modern societies rapidly becoming information societies. As environmental problems, and sometimes environmental catastrophes, inevitably fall upon the modern information society, it is being tested on its capacity to utilize intelligently the masses of information it has the capacity to collect and process. The danger is that our collective efforts to solve environmental problems get trapped in a much-tested cycle: abandoning technically and economically feasible solutions as socially and politically infeasible, yet reacting to failed efforts to solve the problems by amassing more biophysical data on the perceived problems and conducting further technical and economic feasibility studies on them – all with the magnified capacities of modern information technology. The promise of information society is the recognition that the way knowledge is structured into institutionalized rules largely determines our capacity to solve social and environmental problems that are perceived to be urgent. Many of those knowledge structures are embodied in formal institutions, which therefore provide a leverage point for changing the normal but unhelpful ways of doing things.

Yet initiating institutional change is a daunting task. As has been pointed out earlier, the key driver of institutional change over the long run is a discrepancy between informal and formal institutions. I have identified this discrepancy in the four case studies and made an expert assessment of the possibilities for institutional change. In keeping with the role of an expert, I should qualify the analysis by putting it in the context of issues relating to the role of research and expertise in environmental policy. I will first discuss how research methodologies can be adapted to a dynamically evolving institutional setting. Second, I will illustrate the dynamics between research and institutions by describing the reaction that the Californian and Finnish case studies received from the agencies that had ordered the studies. This final chapter concludes with a consideration of the changing role of experts in contemporary environmental management debates.

RESEARCH AND THE EVOLUTION OF ENVIRONMENTAL INSTITUTIONS

Could the idea of organizational uncoupling not have been reached with less sophisticated methods? After all, it has been well documented that many organizations dealing with environmental management have an internal pull between short-term economic interests and long-term environmental ones. And since the idea of institutionalized autonomy is already known and applied in today's societies, could it not have been reached with less legwork and more armchair intelligence? I do not think so.

First of all, while the conflict between short and long time-scales in environmental management is common knowledge in almost any contemporary organization, its cognitive character is not. Much intelligent effort has been devoted to environmental conflicts and their settlement (Amy 1983; Hart *et al.* 1984; Nelkin 1984). But in the case studies in this book, the central long-term environmental policy problem is the absence of a full-blown environmental conflict. Policy makers are professionally convinced that ecological sustainability needs to be addressed in policy, but the conviction is evident only as an unresolved cognitive dissonance, as the circular arguments illustrate. No amount of armchair speculation could have produced this insight. Second, cognitive dissonance is prevalent in any organization and policy regime, and its existence as such does not legitimate a major institutional overhaul (Hosking and Morley 1991). However, in the case study organizations, cognitively dissonant decision makers were found to be part of an organizational culture systemically incapacitated to make decisions. This could not have been discovered without thematic interviews and detailed analysis of individual cognition. Third, while the findings of the four case studies converge on their general themes, the specific processes of institutional–cognitive interaction and the policy impacts derived from them vary from case to case. These details are crucial background information for a more detailed institutional design, and could only be obtained through extensive cognitive mapping.

Questions should also be raised about the possibility of a methodological trap. Was cognitive dissonance found in each case only because of the cognitive mapping approach? This is unlikely, because cognitive mapping does not prejudice the cognitive structures that might emerge in the exercise. In the case studies the structures of interest emerged after careful observation of the data. Causality dominated the data, as one would expect on the basis of cognitive psychology, but other structures, such as the cognitively dissonant issues of the Colorado case, also emerged.

The reforms listed in Table 8.1 contain important lessons on the nature of institutional change. At a glance, the recommendations appear almost paradoxical. The 'easy' reforms of the operational level are often sweeping administrative changes that influence the behaviour of individual decision makers at the highest levels of bureaucracy in profound ways, whereas the 'difficult' reforms of the constitutive level need not change existing administrative structures at all and really have their most significant influence on the small day-to-day decisions of ordinary consumers. Consider Finnish hazardous waste management, which would involve a radical restructuring of the existing semi-governmental hazardous waste monopoly and a major reorientation of the Ministry of the Environment's regulatory policy, if the operational level reform recommendation were to be taken seriously. Contrast that with environmental tax reform, whose most dramatic consequence for individual consumers is that taxes would punish them less for work and more for material consumption and waste.

The paradox that reforms influencing the day-to-day behaviour of individuals are the most significant ones penetrates the very essence of institutions. They are the rules of the social game. The most influential and long-lasting rules are the informal ones, that is, those evident only in the day-to-day thinking and behaviour of individual members of the society. Reforms at the constitutive level attempt to influence the small decisions by individuals, the overarching belief being that the sequence of small decisions by individuals, aggregated over most members of the society, is a significant shaping force for institutional change.

The idea of sequential aggregated decisions acting as the motive force of institutional change invokes notions of uncertainty and complexity. What is there to guarantee that the sum and sequence of individual decisions will have the intended consequences over the long run? Probably nothing. Individuals learn, which means that their day-to-day responses to the institutional rules evolve. Altered individual responses may in turn create political pressures to change the institutional rules themselves. The feedback between formal institutions and individual responses, after all, has historically been at the core of institutional change. While the process may largely fall beyond human control, one can attempt to understand it better through trial and error.

The process of institutional change appears analogous to the non-linear dynamics observed in the so-called chaotic processes. A simple non-linear difference equation describing a complex system, such as the evolution of a species population, displays considerable system stability with certain parameter values, whereas with others it becomes completely unpredictable. But even within the unpredictable range, the unpredictability is patterned in an orderly fashion on recursive runs of the feedback equation. Many systems observed in nature possess non-linear dynamic characteristics, namely they are globally deterministic but locally unpredictable (Dyke 1988; Gleick 1988). To extend the analogy to institutional change, individuals deciding sequentially in accordance with an informal rule are implementing the rule in a recursive feedback. When considered in the aggregate over time, individual decisions take the social system to new unpredictable states. The difference when compared with a natural system is that individuals are cognizant actors capable of reformulating the formal rules they obey in response to the state of the system, as economic historians have observed. The result is a process that has been termed institutional bargaining (Young 1994). Such bargaining can and sometimes does lead to substantial changes in formal institutions. But the battles over proposed changes are likely to be bloody and protracted, and the results are likely to look much more like bargains struck after complex negotiations than elegant blueprints arising from a process of design.

This last point has two important implications for the role of research and expertise in the evolution of environmental institutions. The first has to do with who is considered to be an expert with competence on environmental institutions, the second with the circumstances under which experts can influence the evolution of institutions.

The research approach in each case study in this book was elitist. Snowball sampling as a method embodies the notion that there exists a core group of experts particularly knowledgeable about the issue at hand. An improvement to the method would be to talk to the periphery as well, namely those individuals most people would agree have little say on the outcome of the policy issue, but who are likely to have a profound interest in it. This can be justified on ethical grounds. After all, talking to the fringes of society invokes the Rawlsian notion of maximizing the welfare of the least well-off (Rawls 1971). It can also be justified on policy grounds. A central theme of design for environmental institutions is to create political arenas for local people to influence policy decisions (see Table 9.1). Finally, letting the periphery speak is justified from the theoretical point of view. If one believes that the feedback between mental models and environmental institutions is part of a complex, dynamic and non-linear process, then one should also pay serious attention to the potential for weak signals in the system to trigger significant changes in its overall behaviour. While there is very little we can influence in the system dynamics of social and ecological development, the non-linear dynamics of the system none the less provide chances for triggering change. In the language of non-linear systems, small impulses can push the system from an existing localized equilibrium to new localized equilibria (Clark *et al.* 1995). It is therefore crucially important that we understand the system we are playing with.

To utilize the special features of cognition and self-awareness, social systems should establish monitoring and evaluation systems, to ensure not just technological reliability but also institutional resilience. The reformed constitutive rules, such as environmental taxes or deliberative fora for environmental policy, should be constantly monitored and evaluated. In practice, the monitoring and evaluation requires an expansion of the snapshot-like case studies presented in this book into periodically repeated cognitive mappings of key stakeholder groups, and simultaneous analyses of the relevant institutions. This information should be fed back into the legislative and policy process. Note that this is not a restatement of how the democratic legislative process today operates, but a recommendation to establish a system of institutional observance at a level of social scientific competence and prestige currently devoted only to the development of managerial systems in enterprises aiming at a high level of organizational and technological reliability.

The challenge of institutional reform in environmental policy is the inherent incrementalism and slowness of institutional change. But once the new environmental institutions are in place and perceived to contribute to sustainable development, they are likely to persist by the very fact that they are institutions. Paradoxically, the promise of institutional reform as the guarantor of ecological sustainability lies in the 'rigid technological, organizational, political and attitudinal structures' that the interviewees in the Finnish case study (Tables 6.2 and 6.4) perceived to be at the heart of today's dearth of vision.

To illustrate the institutional rigidities, the following section details the reservations that those who paid for the case studies had about the recommendations. Their reactions are valuable indicators of likely implementation barriers against institutional reform. They also reveal some of the prejudices against the methodologies, which many perceived to be unconventional. It is time for the analyst to lick his wounds.

AGENCY REACTIONS TO REFORM RECOMMENDATIONS

Although this book is primarily about analysing the interaction between expert thinking and environmental institutions, the driving force for the case studies was not analytical but pragmatic interest. The studies were all commissioned and largely funded by governmental agencies. The US Environmental Protection Agency funded the Colorado study; the California Department of Water Resources wanted to know more about the institutions of the state's drainage management; Finland's Ministry of the Environment looked for visions to guide future waste management policies; and the Dutch Development Cooperation Agency facilitated the study on Chinese environmental management. All cared little about how the analyses were conducted, but could not wait to hear how to fix the problem. Chapter 8 described the institutional reform recommendations at the level of detail they were presented to the agencies that had ordered the studies. How successful were the recommendations in inspiring policy action by the agencies?

The short answer is: not very successful. There is an obvious reason for the poor success rate, which is the already-mentioned slowness of institutional change, both the designed and the evolutionary kind. What is more interesting from the policy analytical point of view, however, are the initial reactions to the recommendations from the agencies who had ordered the studies. These reactions tell about the difficulties in launching the institutional changes proposed in Chapter 8. More importantly, they illuminate and reinforce the institutional proposition obtained from the case studies.

I will revisit the Californian and Finnish case studies to illustrate the negative agency reactions to the reform recommendations. In both cases, the decision makers had very good reasons for resisting change, reflecting the resilience of the observed institutional anomalies. Since the toxicity problems in California's irrigation drainage became evident in the early 1980s, there already exists historical evidence of the resilience of the *status quo* that was observed in the case study in the late 1980s: the physical and ecological problems remain virtually unchanged at the time of writing. The Finnish case study is contrasted with an analysis of the institutional constraints of long-term environmental policy and management in the European Union. The discussion is a warning to those optimists who expect far-sighted European environmental policy to result from the replication of national environmental institutions at the transnational level.

More dead ducks in California

When my co-workers and I presented our recommendations to the Californian irrigation agencies in 1988 (Hukkinen *et al.* 1988), we received generally supportive comments. Suggestions and critiques from agency officials had to do with specific points or politically sensitive areas requiring a delicate choice of language. Once language perceived as 'provocative' was altered, the reviews were by and large positive.

However, when we applied for what we thought was an agreed-upon renewal of funding for a second full year of research to focus on more specific data collection, it was denied. We were told that priorities had shifted, that the agencies were not primarily interested in action, that further institutional work was to focus on law, and that there simply was no more funding for work such as ours. According to the officials, our attempts to continue explicating the underlying and urgent dilemma of the irrigation bureaucracy were in fact focused on the 'wrong' problem, because the primary problem remained in the field and not in the bureaucracy. In short, we were trapped by the same dilemma that kept the irrigation bureaucracy shifting the drainage agenda the moment the current one called for policy action.

To find out if the officials' reaction to the institutional reform recommendations had wider support among drainage experts, I sent an article describing the analysis and the recommendations (Hukkinen 1991c) to two members of the US National Research Council's Committee on Irrigation-Induced Water Quality Problems. The Committee had in 1989 published an influential report on nation-wide drainage problems in the US (National Research Council 1989). The members of the Committee responded, both vigorously opposing the idea of uncoupling irrigation from drainage. One felt drainage to be inseparable from irrigation, both philosophically and physically, and that separation would therefore simply lead to absurdities. The other was certain that the new drainage agency would be viewed as just another attempt to provide subsidies to farmers, and that environmental trade-offs made it impossible to manage drainage in an environmentally acceptable manner. No action, combined with more research and public discussion, were in this expert's mind the best options for now.

The reactions from the two experts are understandable, both professionally and politically. In its 1989 report the high-level Committee the experts had belonged to recommended expanding the functions played by the USBR and the local districts to include not just the provision of water for irrigation but also the control and regulation of drainage discharges (National Research Council 1989). Yet one can only wonder why the separation of responsibilities, which one of the experts viewed as absurd, has not led to absurdities in municipal water and waste water management, where the provision of clean water typically belongs to one agency, the consumption of water to independent households and industries, and the treatment of waste water to a third separate agency. Or how any economic activity at all would be possible in the face of negative environmental externalities if, as the other expert stated, environmental trade-offs made environmentally sound drainage categorically impossible. The only certain outcome of wishing for public debate while doing nothing about drainage is more toxicity crises like Kesterson. Public debate, after all, was found to be against the interests of all parties as currently constituted. The prerequisite of an open public drainage discourse over the environmental trade-offs is the clarification of agency agendas by separating the short-term interests of irrigation from the long-term interests of drainage.

In retrospect, the reaction by drainage officials and experts foretold of the persistence of the dilemma they were facing. In 1993 the *San Jose Mercury News*, which is the main newspaper of California's Silicon Valley and one of the most ardent reporters on the Central Valley's environmental problems since the early 1980s, ran a three-article series commemorating the discovery of toxics at Kesterson reservoir a decade earlier (Benson 1993; Thurm 1993a, 1993b). According to the articles, the toxicity problems had not disappeared in ten years. Wildlife biologists interviewed for the articles talked about the 'sons of Kesterson', referring to evaporation ponds of the southern San Joaquin Valley that had over the years since Kesterson proved to be excellent laboratories for studying selenium-induced deformities in birds. The articles were largely based on interviews with representatives from the same interest groups I had interviewed six year earlier, all of whom agreed that the solution, which for them meant long-term maintenance of both salt balance and agricultural production, was no closer than it had been ten years earlier. What is more, they were all resigned to believing that little was going to change in the foreseeable future. An environmental interest group representative felt they had 'lost steam on this one', a farmer irrigating some of the most saline and toxic lands in the valley noted they were 'just marking time here', and the commissioner of US Bureau of Reclamation (USBR) wanted to 'let somebody else jump in front and take the lead' (Thurm 1993a: 29A).

The media have not been the only party to criticize California's irrigation officials for inability to take decisive action on the drainage issue. A report by the Inspector General of the US Department of the Interior, published also in 1993, points out that the USBR spent 50 million US dollars on studying drainage problems without developing a satisfactory management plan. This, the report explains, is due to the fact that the USBR, through the San Joaquin Valley Drainage Program, had persistently pursued the development of in-valley management plans which they knew to be conceptually unworkable or unacceptable to the state of California and the general public (US Department of the Interior 1993).

Agency reactions to our recommendations in 1988 and the worsening drainage problems thereafter are prominent signs of the systemic and resilient character of the drainage dilemma. Acting according to professional convictions about long-term drainage management threatens the political survival of drainage professionals and their agencies, whereas not acting at all threatens the ecological survival of irrigated agriculture as a whole. Not only is the dilemma evident in specific drainage management efforts; it also takes hold of any institutional reform recommendation presented to the irrigation bureaucracy. In the end, the dilemma paralyses the ability of irrigation officials even to think about far-sighted drainage management.

Finnish waste in the European Union

Environmental officials in Finland responded to the recommendations of the study on waste management strategies much as their California colleagues did. Throughout the study, officials at the Finnish Ministry of the Environment, which was the initiator and major supporter of the study, enthusiastically talked about rapid publication of the final report, combined with a public seminar. However, six months after the submission of the final manuscript, the Ministry still had not published the report. In a follow-up meeting I had with the Ministry officials it became evident that, in their view, the diagnosis of environmental corporatism was misconstrued, because corporatist procedures were the way matters had always been dealt with in Finland. The recommendation to uncouple implementation from regulation in waste management was therefore perceived to be of no practical relevance to the Ministry's long-term environmental strategy. After much persuasion, the report was published – without a public seminar. Once again, the institutional recommendations became prisoners of their own logic. The Ministry officials would rather see proposals for Ekokem's management collecting dust on a chief inspector's bookshelf than have them turn long-term hazardous waste management into a potentially controversial public issue.

There are good reasons to believe that the lock-in between expert perceptions and corporatist environmental institutions in Finland is at least as systemic and resilient as the dilemma observed in California. Not only do corporatist structures and procedures exist in other European nation states (such as the Netherlands, as discussed in Chapter 6), but within the bureaucracies of the European Union (EU). Finland joined the EU in 1995, a fact that may potentially reinforce the corporatist institutions already in place in the country.

Recent analyses of the EU and its environmental policy offer telling examples of corporatism, or the institutionalized fusion of conflicting social interests in decision making (Evans *et al.* 1985; Pekkarinen *et al.* 1992; Wilson 1989). The analyses often make no qualitative difference between types of power, but rather talk in purely quantitative terms of an EU institution having too much or too little power (Harrop 1989; Nugent 1989). In environmental policy, the discussion has ranged from power distribution within EU bureaucracy to that between EU and its member states (Wright 1991) to that between EU and other global power blocks (Haigh 1992), but always in purely quantitative terms. Discussion of the qualitative differences and conflicts of interest between implementation and regulation, between different stages of environmental technology, and between short-term economic and long-term ecological concerns, has been absent.

The institutionalized mixing of social interests in Europe is understandable from the historical perspective. EU's decision making institutions are rooted in the European Coal and Steel Community (ECSC), which the Treaty of Paris established in 1951 with a vision of achieving political integration through economic integration. This is reflected in the way power is conceptualized and distributed among EU institutions today.

According to the Rome Treaties of 1957, the Commission proposes, the European Parliament advises, the Council of Ministers decides, and the Court of Justice interprets. The power concept has clearly corporatist characteristics. Europe is assumed to have common economic interests, which are to be pursued through the unanimous project organization of proposing, advising and deciding. To add to the confusion, some of the lines of division over who does what within the EU have become less clear over the past decades. The Council, for example, has usurped some of the Commission's proposing responsibilities by becoming progressively more involved in initiating policy and setting the policy agenda (Nugent 1989).

In practice this means that European-scale proposals on environmental institutions or policy should be critically evaluated to avoid the confusion between different types of power. Dutch policy makers, for example, have expressed their enthusiasm over the target group approach adopted in EU's Fifth Environmental Action Plan, which is modelled after the covenant system of the Netherlands (European Environment Agency 1995; Ministry of Housing, Spatial Planning and the Environment 1994). It should be remembered, however, that covenants (which were discussed in Chapter 6) are an extension of the Dutch procedure for achieving social consensus. They are formulated among a number of parties connected by complex relationships and a long history of negotiation and agreement. In Finland, consensual agreements in environmental policy can similarly be viewed as an extension of the long tradition of labour negotiations between employer and employee unions. But it would be an over-optimistic generalization to assume consensual agreements are feasible at the European transnational scale, where the history, tradition and structure of institutions and organizations vary from country to country. An environmental policy achievable on a European scale would therefore have to be as neutral as possible to the national peculiarities of policy design and implementation.

Economic instruments are one way of levelling the national specifics. While they are not 'neutral' on any single social or cultural scale of measurement, they do provide the advantage of simplifying the terms of international environmental policy debate over equity. Policy debate over Europe-wide environmental taxes, for example, would focus on the equity of distributing a commensurate economic value. In contrast, efforts to establish a transnational negotiation-based regulatory scheme would introduce the complexity of balancing incommensurable values of political debt and commitment between parties from very different national negotiating systems.

Given the EU's historical roots in nation states and its evolution through pragmatic political agreements between them, some analysts feel any democratization of the Union's decision making institutions is highly unlikely (Mann 1993). Policy makers should none the less be aware of the dangers of focusing on the quantitative distribution of power when considering common

European environmental policy. A far more important question is the qualitative distribution of administrative, technological, economic and political authority among existing or completely novel decision making bodies. Due to the large spatial and temporal scales of today's ecological problems, environmental issues will inevitably push more absolute decision making power from individual nation states to the supranational EU. The Single European Act of 1987 recognizes this by making international action an objective of the EU's environmental policy (Duff *et al.* 1994). The globalization of the environmental decision making regime will intensify the sociopolitical divisions that underlie all environmental conflicts, because more nations and generations have vested interests (Hurrell and Kingsbury 1992; World Commission on Environment and Development 1990). Corporatist institutions lack the transparency required of a decision making system aiming to settle the long-term, multi-dimensional issues of a transnational environment (Hukkinen 1995c).

When the Finns were still pondering whether to join the EU, some environmentalists in the country were optimistic that the EU would give a supranational boost to environmental policies at the national level. While the political clout of the EU on national level affairs is indisputable, it is not so clear that the presence of a supranational power as such would ensure open debate and resolution of environmental conflicts. On the contrary, environmental institutions of the EU as currently configured would be likely to prolong and strengthen the systemic inability of Finnish society to make long-term environmental policy.

PUBLIC LIFE AND THE POLITICS OF EXPERTISE

Negative agency reactions to the institutional reform recommendations do carry a positive message. First, the reactions verify the main institutional proposition of this study concerning the mutually reinforcing lock-in between expert models and environmental institutions. A verified hypothesis is always gratifying to a researcher, even if it means discontinuation of further research. Second, since institutional change is known to be slow, there is reason to be optimistic that at some point in the future the recommendations will resonate with the mental models of the policy makers better than they did at the time of the studies. Finally, the very inertia of all institutions bodes well for the endurance of institutions of sustainability, once they do emerge.

There is a strategic lesson to be learned from agency reactions as well. Anyone wishing to set institutional reform in motion needs to be profoundly aware that the lock-in between cognitively dissonant mental models and environmental institutions is just as nested as the institutions themselves. Administrative uncoupling of implementation and regulation at the operational level, for example, can only start when the impulse for change comes from outside the range of applicability of operational rules. In practice, the initiator of institutional reform should have powers either at a deeper level of rules or within a regime perceived to lie beyond environmental policy and management.

A healthy dose of scepticism is also called for when considering the problem analyses and policy recommendations of the case studies. I cannot empirically argue that the decision makers who rejected the analyses were 'wrong'. And I have no empirical grounds to argue that following the policy recommendations would have produced 'better' outcomes than the *status quo*. I can only base my argument on what institutional analysts have learned of the dynamics between individuals and institutions, often in regimes other than environmental management. As a policy analyst, I would like to be able to say something more than simply stating that I and, in the Californian case study, my co-workers had what looked like a good idea and that the officials rejected it because it was not in their problem frame. In other words, I would like to close with something more action-oriented than the diagnosis of a circular problem that replicates itself in the decision makers' reaction to the analysis.

There is something experts can do now, instead of just waiting for the institutions to grow into accepting their recommendations. It is what I as an expert have done in writing this book, namely recounted in public the experiences of the four case studies. In doing so, I have, using Beck's (1992) terminology, entered the sub-politics of environmental management. In other words, while making an effort to open up political discourse on the structure and functioning of environmental institutions, I have at the same time raised broader questions about rights, responsibilities, and power relationships in society. I and my co-workers have transcended the interests of those who ordered the studies through a transformation from experts in an expert system into participants in a democratically dialogical public sphere (Beck *et al.* 1995). The case studies illustrate this process. Each study started out with the recognition of a problematic situation within the regime of environmental management. But as the analyses unravelled the nested structure of environmental institutions, the policy recommendations inevitably expanded from the operational details of technological reliability and regulatory responsibility to the constitutive issues concerning the relationship between short-run economic and long-run environmental conditions.

While the publication of research results is a far cry from a scientifically enlightened political process, such as the participatory environmental assessment envisioned in Chapter 8, publicity as such is a key ingredient in successful forcing of issues from the realm of technology management into the realm of environmental sub-politics (Beck 1992). In fact, future political debates over sustainable development may well be such that, apart from well-crafted economic instruments, publicity is the only currency with which an agency with a long-run sustainability mandate can counter powerful players with short-run economic interests. In the Finnish case study, for example, the persuasion that was required in order for the Ministry to publish the report on the long-term strategies of Finland's waste management was a letter I wrote to the Environment Minister expressing my concern over the fact that the Ministry's officials were dragging their feet with the publication of the final report. After a week I received a phone call from the Ministry's publications editor requesting two figures for the report, which was otherwise ready for printing.

These are endless games. Yet playing the system should come quite naturally to us. From birth we have learned the rules of the game through trial and error, and continue learning throughout our lives. That is what my daughter was doing with Santa Claus on Christmas Eve in 1996. She was cunning, too, in response to all those pre-Christmas threats, treating with respect the formal institution when it was present, yet voicing strong reservations about it in its absence. In a couple of years, she will have abandoned her belief in that particular institution. Similar abandonment of old institutions would serve us well in our search for novel configurations of social rules that would enable us to reflect – and decide – on our common environment over the long run.

APPENDIX 1

EXAMPLES OF THE CODIFICATION OF INTERVIEW TRANSCRIPTS INTO ISSUE AND PROBLEM NETWORKS

The following two examples illustrate the transformations that took place in the codification of narrative transcripts into issue and problem networks. The first example presents the formulation of issue 44 in the Colorado case study (Chapter 4). Issue 44 plays a central role in the main argument of the study. It describes how the Interactive Accounting Model (IAM), or the Burns model, is being used as a cheap, and from the modelling point of view inappropriate, replacement for detailed monitoring and evaluation of actual water quality, so as to obtain regulatory acceptance and funding for further on-farm management experiments. The second example details the construction of the irrigation bureaucracy's terminal loop in the Californian case study (Chapter 5). The terminal loop describes how the retirement of agricultural land as a result of drainage problems and the conflicts among irrigators over scarce resources, profits and subsidies reinforce each other, and gradually diminish the viability of irrigated agriculture in California's Central Valley.

CORRESPONDENCE BETWEEN ISSUE 44 AND INTERVIEW TRANSCRIPTS IN THE COLORADO CASE STUDY

Issue 44

As a lumped parameter mass balance model, the Burns model is incapable of accurately predicting the impact of on-farm management practices on water quality. But the SCS (Soil Conservation Service) only needed a crude basin-wide planning model to quantify the long-term effects of the Patterson Hollow project on off-site water users and to justify the project to funding agencies.

APPENDIX 1

Supporting transcripts

We [the SCS] told them [the local farmers] there's some government money to improve their system, and they'd go, 'how can we get it?' you know. We said sure, we'll look at it, as soon as we have time. Part of the project development, of course, is what's going to be the effect of it. Fairly easily we can say that if I can increase the consumptive use, that will increase the production levels on farm. We have techniques to do that. But I cannot answer the question what effect that might have on the water quality. Water quality in Las Animas is around 3,000 ppm. Certainly that exceeds your drinking water standards, fish standards, and all kinds of things. If you build this project here, will it lower the concentration enough to meet the drinking water standards? I have no idea. And that is what we're trying to use the [Burns] model to display. We're not so concerned about an absolute value. But I have a reasonable indication that I'm going to have a 30 per cent reduction maybe. Or it may show that our work here may not be significant. For our work, we can live with those kinds of answers. If you worked for the State Department of Health's Water Quality Division, it might not be an adequate answer. I don't think this model could help, if somebody came and said they wanted to have the TDS [total dissolved solids] at some specific level.

(interview 1)

In my mind, because you're dealing with a canal as a unit, and because there isn't a mechanism within the [Burns] model to look at different irrigation practices – and those are lumped into a fairly small number of parameters – it seems that the value of the model is probably more in looking at the big picture rather than looking at individual on-farm changes . . . I don't know all the details of the history of [the SCS] project, what kind of timeline they had, and what kind of internal review. But I remember [a representative of the SCS] saying specifically that for them to get a project off the ground they needed to have some mechanism that they could show benefits. It seemed that the model gave them some mechanism to do that. It might help them justify the project to be able to say that salinity in, say, Las Animas would decrease by some percentage.

(interview 2)

Yes, we're [the SCS] going to be using the [Burns] model there. When water quality became the buzzword, and the SCS was being directed to look at water quality as a programme, why, we said

OK, we need to be generating projects – on-farm projects to address water quality problems, salinity being one of those problems. Initially, the only mode through which we could do those kind of projects was through PL-566 projects, watershed projects. We wrote proposals for several PL-566 projects . . . Then we got directions back from our Tech Center in Washington saying, well, these look OK. But you're going to have to do a much better job in justifying them, based not on the merits within the boundary of the watershed but what effects is it going to have outside of it. We were told that the political trend in Washington is to fund projects that have off-site benefits rather than within the watershed itself. So we said, golly, how do we do that? [Explains how the US Geological Survey provided the SCS with the Burns model.] We then had a model to play what-if games. But then the Patterson Hollow hydrologic unit came along. And of course over the years they had been beating it into our heads that the SCS is not a monitoring agency. After we got back the comments of the Patterson Hollow acceptance, they said, we accept your proposal, we're going to fund it. However, you need to do monitoring and evaluation. We only asked for *x* amount of dollars, which did not include money for staff and equipment to do monitoring and evaluation. My task became then to come up with a plan to fulfil our obligations for monitoring and evaluation, but without any money. I said, I'm going to take the path of least resistance. I'm going to take the Burns model to show what wonderful things we've done by spending 2.5 million dollars in the hydrologic unit on on-farm practices. Having nice computer printouts and graphics should impress the people in Washington who don't know any better.

(interview 3)

On a very gross scale [the Burns model can make the linkage between management practices in the field and the water quality in the river]. They're [the SCS] going to have to be very careful, because we more or less lumped the entire area under a canal. They're going to have to do some work to translate what their individual field practices would do even under that whole canal, and some of those canals are large. It will never give them a calibrated tool. All it will give them is a planning tool to help integrate the individual impacts they have into a basin. And there I think it can be a useful tool. But I hope they won't try to go in and, say, calibrate on individual ditches or anything, because they'll lose it. It's just not that scale of a product.

(interview 7)

CORRESPONDENCE BETWEEN THE IRRIGATION BUREAUCRACY'S TERMINAL LOOP AND INTERVIEW TRANSCRIPTS IN THE CALIFORNIAN CASE STUDY

The construction of the terminal loop will be illustrated by first presenting the two transcripts from which the loop was constructed. Thereafter transcripts will be quoted that were coded as 'tangential' to the feedback loop, i.e. sharing at least one node with the loop.

The terminal loop

(8.4.0) Equity problems created by cost-sharing arrangements of drainage management solutions: who benefits, who pays?
→(5.3.3) Business competition in agricultural community.
→(6.3.0) Large farms versus small farms.
→(7.2.5) Land must go out of production.
→(8.5.0) Need for compensation for farmers going out of business due to drainage problems.
→(8.4.0)

Transcripts supporting the loop

(8.4.0)→(5.3.3)→(6.3.0)→(7.2.5):

> In the '60's the two main disposal options for drainage were either the Bay or the ocean at Monterey or Morro Bay. The probable course of events today is that some farmers who have marginal agricultural lands cannot bear more costs and will be driven out of business. Cropping patterns have changed and will be changing. Since it is a benefit to all members of the society to have farmers doing OK, some support to the drainage problem solutions should come from taxpayers. It is an analogous situation to the military being supported by taxes. However, we have the kind of government that emphasizes individual choice, so this kind of support is hard to come up with today. Therefore, some lands will probably be going out of production. Small farmers are the first ones to drop out, because they are traditionally poor businessmen. This is too bad, because it is poor management of land and water resources. Ideally, one could think of some kind of subsidy or incentive programme to solve the problem.
>
> (interview 1)

(7.2.5)→(8.5.0)→(8.4.0):

> The main economic questions pertinent to the problem include: is

> it affordable and economically feasible to solve the problem and continue agriculture, or will the only affordable solution be to stop irrigating the lands that cause the problem? What can the farmers pay to solve the problem? Should the society pay a portion of the solution? Should actions be directed toward a final solution, which would both preserve the agricultural productivity and enhance the wetlands habitat? Sociopolitical problems at the state level stem from the link between water exports and agriculture. Local socio-economic questions include compensation for those farmers who cannot irrigate anymore due to the drainage problems.
>
> (interview 3)

Transcripts supporting problem networks tangential to the loop

(5.3.3) Business competition in agricultural community.
→(8.4.0) Equity problems created by cost-sharing arrangements of drainage management solutions: who benefits, who pays?

> The perspective has not changed much since then. It's a 'dog eat dog' world, a business world, that exists between the farmers of the valley. The farmers are fighting each other and are not willing to make sacrifices. The problem, therefore, is one of equity: first, how to find an acceptable definition of equity; second, how to come up with a mechanism to solve the problem of equity.
>
> (interview 5)

(6.3.0) Large farms versus small farms.

> There are problems in the relationship between the Grasslands Basin Districts and Westlands Water District. The Grasslands Basin Districts are older water rights holders, especially the exchange contractors. They have a different posture than Westlands does. The landowners are smaller than those in Westlands. So there is a difference both in timing and size. The San Luis Unit also triggered disputes between the Grasslands Basin and Westlands. Westlands is a well organized entity, whereas the Grasslands Basin Districts tend to fight also within themselves.
>
> (interview 15)

(8.2.0) Water discharged to sea is not water wasted, and should be given a value in economic analyses.
→(8.9.0) Agricultural drainage costs are not internalized in farming to make farmers realize that the wrong kind of land is in irrigated agriculture.
→(7.2.5) Land must go out of production.

> The problem is that there are entrenched financial and political interests in agricultural drainage that need to be changed. The hardest things to change are the practices of the farmers. There is still an ideology of frontierism among them, with concepts of unlimited land and water resources. Kesterson triggered the latest disputes. A lot of public dollars would be saved if the bad farming and irrigation practices were stopped. The craziness in California is that subsidized cotton is being irrigated with subsidized water. The Delta ties in with drainage questions because a lot of the drainage waters are being discharged into the San Joaquin River, which ends up in the Delta. If the Delta were to be evaluated, a value should be put on the water being wasted to the sea. This is not accepted in the current economic analyses . . . Irrigation of farmlands with drainage problems should be stopped. Water saved should be directed to the urban areas to let them grow and eventually die off in their pollution. The new axis of water disputes in California is not North vs South, but rather Agricultural vs Urban. The idiocy of throwing away good water, as is being done in the valley now by the farmers, will probably stop.
>
> (interview 20)

(7.2.5) Land must go out of production.
→(1.9.0) Agricultural drainage problems may take farmers out of production, which means no payers for the federal and state water projects.

> DWR and USBR have a definite concern in agricultural drainage. Their worry is that drainage problems will take farmers out of production, which would mean no payers for their water supply.
>
> (interview 8)

(8.6.0) Hardships on agricultural community resulting from combined effects of weak agricultural economy and drainage problems.
→(7.2.5) Land must go out of production.

> There has been a drop in the farmed acreage of land in this area. In 1974 it was about 970,000 acres, and has dropped since 160,000 acres. This is due to the cotton market prices. The drop in acreage has occurred in the poor quality lands, i.e. those lands subject to agricultural drainage problems . . . The chain has been from vegetables to cotton to taking land out of production . . . Commodity prices are more likely to affect the drainage problem development here. When commodity prices fall, as has happened in the recent years, it is the poor quality land with drainage problems that will first go out of production.
>
> (interview 19)

(7.2.5) Land must go out of production.

> Within the local districts there is little conflict because farmers have the same goal, i.e. keeping land in production . . . EDF's [Environmental Defense Fund] position in the drainage problem is in concert with ours. They want something to be done through in-valley treatment and disposal. They do not want land out of production, and are contradicting the NRDC's [Natural Resources Defense Council] position in this issue.
>
> (interview 21)

APPENDIX 2

PRESENTATION OF AGGREGATED PROBLEM NETWORKS AS BAYESIAN NETWORKS

The fishnet-like drainage problem networks described in the Californian case study (see Chapter 5) have two structural characteristics – single connectedness and cycles – which demand special attention in the design of the computational architecture.

BELIEF PROPAGATION IN SINGLY CONNECTED NETWORKS

In singly connected networks, unlike tree-structured networks, a node may have multiple parents, which permits 'sideways' interactions via common descendants (Figure A1). If a node has several parents, the summation over all value combinations of parent variables becomes too tedious, requiring approximation techniques that utilize the structure of the conditional link probability matrix $P(x|u_1, \ldots, u_n)$. Using the so-called disjunctive interaction model, belief propagation equations for the singly connected network (equations A1 through A4) shown in Figure A1 are as follows (for details, see Pearl 1986, 1988).

Belief in (or probability of) problem statement X in Figure A1, given the probability of the rest of the network, is expressed as the dynamic node probability, $BEL(x)$:

$$BEL(x) = \begin{cases} \alpha\lambda_0 \prod_i (1 - c_i \pi_{iX}) & \text{if } x = 0 \\ \alpha\lambda_1 [1 - \prod_i (1 - c_i \pi_{iX})] & \text{if } x = 1 \end{cases} \qquad \text{(A1)}$$

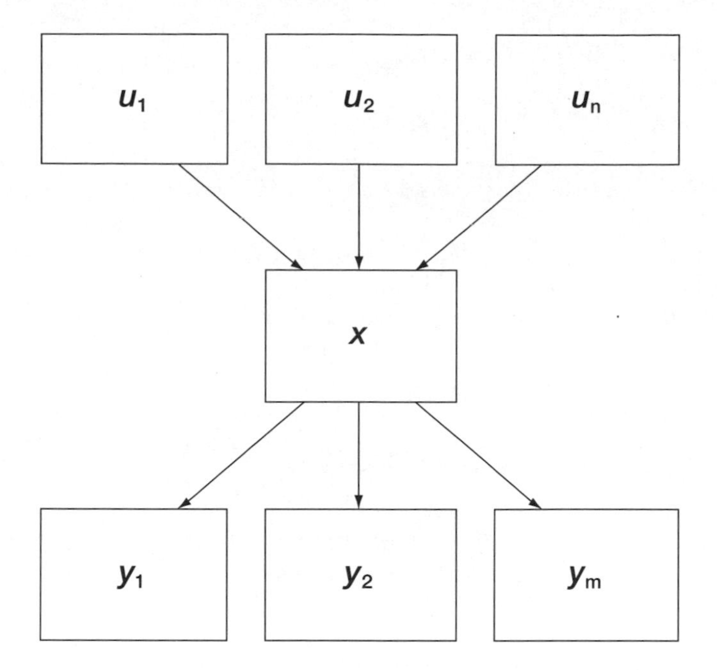

Figure A1 A singly connected Bayesian network.

where x stands for the values the descendant node X can receive ($x = 0$ and $x = 1$ represent the events X = FALSE and X = TRUE, respectively); $\lambda(x) = (\lambda_0, \lambda_1)$ represents the combined evidential support contributed by X's children; π_{iX} is the causal support from a TRUE parent, U_i = TRUE; c_i represents the degree to which an isolated explanation U_i= TRUE can endorse the consequent event X = TRUE; and α is a normalizing constant. Bottom-up propagation of belief occurs through $\lambda_X(u_i)$, which is the new message that node X sends to its parents U_i:

$$\lambda_X(u_i) = \begin{cases} \lambda_0 q_i \Pi_i' + \lambda_1(1 - q_i \Pi_i') & \text{if } u_i = \text{TRUE} \\ \lambda_0 \Pi_i' + \lambda_1(1 - \Pi_i') & \text{if } u_i = \text{FALSE} \end{cases} \quad \text{(A2)}$$

where

$$\Pi_i' = \prod_{k \neq i} (1 - c_k \pi_{kX}) \quad \text{(A3)}$$

and q_i denotes the probability that the i-th parent of X is FALSE (i.e. $c_i = 1 - q_i$). Finally, in top-down propagation the new $\pi_{Y_j}(x)$ message from X to its descendant Y_j is calculated by:

$$\pi_{Y_j}(x) = \begin{cases} \prod_{k \neq j} \lambda_{Y_k}(x) \left[1 - \prod_i (1 - c_i \pi_{iX})\right] & \text{if } x = 1 \\ \prod_{k \neq j} \lambda_{Y_k}(x) \prod_i (1 - c_i \pi_{iX}) & \text{if } x = 0 \end{cases} \quad \text{(A4)}$$

The boundary conditions for a singly connected network are as follows.

1 Initial nodes: if X is a node with no parents, $\pi(x)$ is equal to the prior probability. In drainage problem networks, the prior probability is determined by the number of times a problem statement has been mentioned by experts, normalized so that the prior probabilities of all 90 problem statements sum up to unity.
2 Terminal nodes: if X is a node without descendants and it has not been given a prior value to deal with cycles (see following section), then $\lambda(x) = (1,1, \ldots, 1)$.
3 Evidence nodes: if evidence X = TRUE(FALSE) arrives (X being any node in the network), then $\lambda(x) = 1(0)$. For reasons stated below, loops are treated as true evidence nodes in drainage problem networks.

DEALING WITH CYCLES

Both directed and undirected cycles are problematic in Bayesian belief propagation. Directed cycles, or loops, are not allowed in Bayesian networks, and undirected cycles enable messages in the network to circulate indefinitely around the cycles, whereby the process may not converge to a stable equilibrium (Pearl 1988).

The directed cycles in drainage problem networks are positive feedback loops in which problems reinforce themselves. In probabilistic terms, such a network configuration increases the likelihood of the loop problems indefinitely, making their probabilities approach unity. To do away with the inadmissible loop structure, the individual problem statements of a loop are first clustered into a single problem node (Pearl 1988), which is then given the probability 1 before belief propagation. After clustering, incoming links to any problem node in the loop are treated as incoming links to the 'loop node', and outgoing links are interpreted similarly.

A network with undirected cycles, on the other hand, is made singly connected through conditioning, i.e. by first assuming that a selected group of variables (called the cycle cut-set) has zero probability, then assuming the variables have a probability of one, and finally averaging the two results (Pearl 1988). Cycle cut-set is the set of nodes that breaks a cycle in a network and renders it singly connected (Geffner and Pearl 1987; on cut-set in general, see Harary *et al.* 1965; Robinson and Foulds 1980; Wilson 1974).

BIBLIOGRAPHY

Abbott, P.O. (1985) *Description of Water-Systems Operations in the Arkansas River Basin, Colorado*, US Geological Survey Water-Resources Investigations Report 85–4092, Lakewood, CO: US Geological Survey.

Allenby, B.R. and Richards, D.J. (eds) (1994) *The Greening of Industrial Ecosystems*, Washington, DC: National Academy Press.

American Heritage Dictionary (1986) based on the New Second College ed., New York: Dell Publishing.

Amy, D.J. (1983) 'The politics of environmental mediation', *Ecology Law Quarterly*, 11, 1: 1–19.

Arad, N. and Glueckstern, P. (1981) 'Desalination, a review of technology and cost estimates', in D. Yaron (ed.), *Salinity in Irrigation and Water Resources*, New York: Marcel Dekker, pp. 325–61.

Arctic Environmental Protection Strategy (1997) *Guidelines for Environmental Impact Assessment (EIA) in the Arctic*, Helsinki: Finnish Ministry of the Environment.

Assmuth, T., Poutanen, H., Strandberg, T., Melanen, M., Penttilä, S. and Kalevi, K. (1990) *Environmental Impacts of Hazardous Waste in Landfills* (Kaatopaikkojen ongelmajätteiden ympäristövaikutukset), in Finnish with English summary, Series A: 67, Helsinki: National Board of Waters and the Environment.

Ausubel, J.H. and Sladovich, H.E. (eds) (1989) *Technology and Environment*, Washington, DC: National Academy Press.

Ayres, R.U. (1989) 'Industrial metabolism', in J.H. Ausubel and H.E. Sladovich (eds) *Technology and Environment*, Washington, DC: National Academy Press, pp. 23–49.

Ayres, R.U. and Kneese, A.V. (1969) 'Production, consumption, and externalities', *American Economic Review*, 59: 282–97.

Ayres, R.U. and Kneese, A.V. (1989) 'Externalities: economics and thermodynamics', in F. Archibugi and P. Nijkamp (eds) *Economy and Ecology: Towards Sustainable Development*, Dordrecht: Kluwer, pp. 89–118.

Babbie, E. (1986) *The Practice of Social Research*, 4th ed., Belmont, CA: Wadsworth Publishing.

Bailey, K.D. (1982) *Methods of Social Research*, 2nd ed., New York: The Free Press.

Beaugrande, R. de (1980) *Text, Discourse, and Process: Toward A Multidisciplinary Science of Texts*, Vol. IV in Advances in Discourse Processes, Norwood, NJ: Ablex Publishing.

Beaumont, J.R., Pedersen, L.M. and Whitaker, B.D. (1993) *Managing the Environment: Business Opportunity and Responsibility*, Oxford: Butterworth Heinemann.

Beck, U. (1992) *Risk Society: Towards a New Modernity*, trans. M. Ritter, London: Sage.

Beck, U., Giddens, A. and Lash, S. (1995) *Reflexive Modernization: Politics, Tradition and Aesthetics in the Modern Social Order*, Cambridge: Polity Press.

Bell, D. (1967) 'Working session I: a summary by the chairman', *Daedalus* (Toward the Year 2000: Work in Progress), summer 1967, pp. 698–704.

Bennett, J.W. (1996) *Human Ecology as Human Behavior: Essays in Environmental and Development Anthropology*, expanded ed., New Brunswick: Transaction Publishers.

Benson, M. (1993) 'Kesterson's human toll', *San Jose Mercury News*, 5 July, pp. 1A, 16A.

Berger, P.L. and Luckmann, T. (1967) *The Social Construction of Reality: A Treatise in the Sociology of Knowledge*, New York: Doubleday.

Bertalanffy, L. von (1968) *General System Theory: Foundations, Development, Applications*, New York: George Braziller.

Blomquist, W. (1992) *Dividing the Waters: Governing Groundwater in Southern California*, San Francisco, CA: ICS Press.

Boer, M. de (1995) 'Opening address', in S. Zwerver, R.S.A.R. van Rompaey, M.T.J. Kok and M.M. Berk (eds), *Climate Change Research: Evaluation and Policy Implications*, Proceedings of the International Climate Change Research Conference, Maastricht, The Netherlands, 6–9 December 1994, Amsterdam: Elsevier.

Bower, G.H. and Morrow, D.G. (1990) 'Mental models in narrative comprehension', *Science*, 247: 44–8.

Bramer, M.A. (1985) 'Expert systems: the vision and the reality', in M.A. Bramer (ed.), *Research and Development in Expert Systems*, Proceedings of the Fourth Technical Conference of the British Computer Society Specialist Group on Expert Systems, University of Warwick, 18–20 December 1984, Cambridge: Cambridge University Press, pp. 1–12.

Buchholz, R.A. (1993) *Principles of Environmental Management: The Greening of Business*, Englewood Cliffs, NJ: Prentice Hall.

Burns, A.W. (1988) *Computer-Program Documentation of an Interactive–Accounting Model to Simulate Streamflow, Water Quality, and Water-Supply Operations in a River Basin*, US Geological Survey Water-Resources Investigations Report 88–4012, Denver, CO: US Geological Survey.

Burns, A.W. (1989) *Calibration and Use of an Interactive–Accounting Model to Simulate Dissolved Solids, Streamflow, and Water-Supply Operations in the Arkansas River Basin, Colorado*, US Geological Survey Water-Resources Investigations Report 88–4214, Lakewood, CO: US Geological Survey.

Burt, R.S. (1982) *Toward A Structural Theory of Action: Network Models of Social Structure, Perception, and Action*, New York: Academic Press.

Cain, D. (1985) *Quality of the Arkansas River and Irrigation-Return Flows in the Lower Arkansas River Valley, Colorado*, US Geological Survey Water-Resources Investigations Report 84–4273, Lakewood, CO: US Geological Survey.

Cain, D. (1987) *Relations of Specific Conductance to Streamflow and Selected Water-Quality Characteristics of the Arkansas River Basin, Colorado*, US Geological Survey

Water-Resources Investigations Report 87–4041, Denver, CO: US Geological Survey.
Cairncross, F. (1991) *Costing the Earth*, London: Economist Books.
California Department of Public Works (1930) *Report to Legislature of 1931 on State Water Plan*, Bulletin No. 25, Sacramento, CA: California State Printing Office.
California Department of Public Works (1931) *San Joaquin River Basin*, Bulletin No. 29, Sacramento, CA: Department of Public Works, Division of Water Resources.
California Department of Water Resources (1957) *The California Water Plan*, Bulletin No. 3, Sacramento, CA: Department of Water Resources, Division of Resources Planning.
California Department of Water Resources (1965) *San Joaquin Valley Drainage Investigation: San Joaquin Master Drain*, Bulletin No. 127, preliminary ed., January, Sacramento, CA: Department of Water Resources.
California Department of Water Resources (1974), *Status of San Joaquin Valley Drainage Problems*, Bulletin No. 127–74, Sacramento, CA: Department of Water Resources.
California Department of Water Resources (1987) *California Water: Looking to the Future*, Bulletin 160–87, Sacramento, CA: Department of Water Resources.
California Senate Permanent Fact Finding Committee on Water Resources (1965) *Progress Report to the Legislature, 91965 Regular Session*, Sacramento, CA: Senate of the State of California.
California State Water Resources Control Board (1986) *Workplan for San Joaquin River Basin Water Quality Objectives and Regulation of Agricultural Drainage Discharges*, SWRCB Order No. W.Q. 85–1 Technical Committee, Sacramento, CA: State Water Resources Control Board.
California State Water Resources Control Board (1987) *Regulation of Agricultural Drainage to the San Joaquin River*, SWRCB Order No. W.Q. 85–1 Technical Committee Report, draft, May, Sacramento, CA: State Water Resources Control Board.
California State Water Resources Control Board (1988) *Executive Summary: Water Quality Control Plan for Salinity, San Francisco Bay/Sacramento–San Joaquin Delta Estuary*, draft, October, Sacramento, CA: State Water Resources Control Board.
Chinese–English Dictionary (1985) Beijing: Beijing Foreign Language College.
Chinese Management Science Dictionary (1985) Taipei: Guo Li Zhong Xin University.
Christiansen, F. and Rai, S.M. (1996) *Chinese Politics and Society: An Introduction*, London: Prentice Hall.
Churchman, C.W. (1983) *The Systems Approach*, revised and updated, New York: Dell.
Clark, C.H. (1958) *Brainstorming: The Dynamic New Way to Create Successful Ideas*, Garden City, NY: Doubleday & Co.
Clark, N., Perez-Trejo, F. and Allen, P. (1995) *Evolutionary Dynamics and Sustainable Development: A Systems Approach*, Aldershot: Edward Elgar.
Commissioners for Colorado (1948) *Report and Submission of the Arkansas River Compact*, negotiated and signed by Commissioners representing the states of Colorado and Kansas at Denver, Colorado, 14 December, 1948, Denver, CO: State of Colorado.
Constanza, R., Daly, H., Hawken, P. and Woodwell, J. (1995) 'Non-partisan ecological tax reform: a win–win proposal that is economically efficient, socially

equitable, and ecologically sustainable', *International Society for Ecological Economics Newsletter* 6, 3: 3, 8.

Côté, R.P., Ellison, R., Grant, J., Hall, J., Klynstra, P., Martin, M. and Wade, P. (1994) *Designing and Operating Industrial Parks as Ecosystems*, Dalhousie: Dalhousie University, School for Resource and Environmental Studies.

Cottrell, R. (1995) 'A vacancy awaits', *The Economist: A Survey of China*, 18 March, pp. 1–26.

Dai Yiao Nan and Zhang Xi Hen (eds) (1984) *Environmental Protection Worker Practical Handbook* (Huan Jing Bao Hu Gong Ren Shi Yong Shou Ce), Beijing: Metallurgical Industry Press.

Dijk, T. van (1977) 'Connectives in text grammar and text logic', in T.A. van Dijk and J. Petöfi (eds), *Grammars and Descriptions*, Research in Text Theory, Vol. 1, Berlin: Walter de Gruyter, pp. 11–63.

Diringer, E. (1988a) 'High levels of selenium in nature areas', *San Francisco Chronicle*, 11 May, p. A1.

Diringer, E. (1988b) 'Hodel move snags Kesterson cleanup', *San Francisco Chronicle*, 16 April, p. A13.

Diringer, E. (1989) 'Warning of new toxic spots: "final" Kesterson cleanup plan OKd', *San Francisco Chronicle*, 22 September, p. A4.

Douglas, M. and Wildavsky, A. (1983) *Risk and Culture: An Essay on the Selection of Technical and Environmental Dangers*, Berkeley: University of California Press.

Dovers, S.R. (1995) 'A framework for scaling and framing policy problems in sustainability', *Ecological Economics*, 12, 2: 93–106.

Dryzek, J.S. (1997) *The Politics of the Earth: Environmental Discourses*, Oxford, Oxford University Press.

Duda, R.O., Hart, P.E. and Nilsson, N.J. (1976) 'Subjective Bayesian methods for rule-based inference systems', in *Proceedings of 1976 AFIPS* (American Federation of Information Processing Societies) National Computer Conference, New York City, 7–10 June, Montvale, NJ: AFIPS Press, pp. 1075–82.

Duff, A., Pinder, J. and Pryce, R. (eds) (1994) *Maastricht and Beyond: Building the European Union*, London: Routledge.

Dunning, H.C. (1982) *Water Allocation in California: Legal Rights and Reform Needs*, Institute of Governmental Studies Research Paper, Berkeley, CA: Institute of Governmental Studies.

Dyke, C. (1988) *The Evolutionary Dynamics of Complex Systems: A Study in Biosocial Complexity*, New York: Oxford University Press.

Eckhouse, J. (1989) 'Farmers face perilous times', *San Francisco Chronicle*, 13 March, pp. C1, C3–C4, C8.

Economist, The (1991) 'America's electricity industry: less is more', 26 October, p. 87.

Economist, The (1992) 'China: breathing the air of success', 15 February, p. 62.

Economist, The (1995) 'Climate change: smoke', 8 April, p. 58.

Eden, C. (ed.) (1992) *Journal of Management Studies: Special issue on cognitive maps*, Vol. 29, No. 3.

Eden, C., Jones, S. and Sims, D. (1979) *Thinking in Organizations*, London: Macmillan Press.

EDF Letter (1987) 'California agency adopts EDF plan to solve toxic wastewater problem', 18, 5: 1, 8.

Elliott, C.G. (1904) *Drainage of Farm Lands*, US Department of Agriculture Farmers' Bulletin No. 187, Washington, DC: Government Printing Office.
Engelbert, E.A. and Munro, J.F. (1982) *Incentives and Disincentives for Water Conservation in the California State Water Policymaking Process*, United States of America: US Department of the Interior, Office of Water Research and Technology.
Environment Yearbook of China (Zhong Guo Huan Jing Nian Jian) (1993) Beijing: Metallurgical Industry Press.
EPOC AG (1987) *Removal of Selenium from Subsurface Agricultural Drainage by An Anaerobic Bacterial Process*, Final Report to California Department of Water Resources, Fresno, CA: EPOC AG.
European Environment Agency (1995) *Environment in the European Union – 1995: Report for the Review of the Fifth Environmental Action Programme*, Luxembourg: Office for Official Publications of the European Communities.
Evans, P.B., Rueschemeyer, D. and Skocpol, T. (eds) (1985), *Bringing the State Back In*, Cambridge: Cambridge University Press.
Fairclough, J., Whelan, B. and Williams, J. (1993) 'The business of green design', *Greener Management International*, No. 2, April, pp. 14–20.
Fay, J.S., Fay, S.W. and Boehm, R.J. (1991) *California Almanac*, 5th ed., Santa Barbara, CA: Pacific Data Resources.
Festinger, L. (1962) [1957] *A Theory of Cognitive Dissonance*, Stanford, CA: Stanford University Press.
Fielding, N.G. and Fielding, J.L. (1986) *Linking Data*, Qualitative Research Methods, Vol. 4, Beverly Hills, CA: Sage.
Forsyth, R. (1984) 'The architecture of expert systems', in R. Forsyth (ed.), *Expert Systems: Principles and Case Studies*, London: Chapman and Hall, pp. 9–17.
Fortier, S. and Cone, V.M. (1909) *Drainage of Irrigated Lands in the San Joaquin Valley, California*, US Department of Agriculture, Office of Experiment Stations, Bulletin 217, Washington, DC: Government Printing Office.
Frosch, R.A. and Gallopoulos, N.E. (1989) 'Strategies for manufacturing', *Scientific American*, 261, 3: 144–52.
Fujii, R. (1988) *Water-Quality and Sediment-Chemistry Data of Drain Water and Evaporation Ponds from Tulare Lake Drainage District, Kings County, California, March 1985 to March 1986*, US Geological Survey Open-File Report 87–700, Sacramento, CA: US Geological Survey.
Geffner, H. and Pearl, J. (1987) 'Distributed diagnosis of systems with multiple faults', in Institute of Electrical and Electronics Engineers, Inc. (IEEE), *Proceedings of the Third Conference on Artificial Intelligence Applications*, Kissimmee, FL, 23–7 February, Washington, DC: IEEE Computer Society Press, pp. 156–62.
Giddens, A. (1986) *The Constitution of Society: Outline of the Theory of Structuration*, Berkeley: University of California Press.
Gleick, J. (1988) *Chaos: Making a New Science*, New York: Penguin Books.
Godet, M. (1993) *From Anticipation to Action: A Handbook of Strategic Prospective*, trans. C. Degenhardt, revised V. Shepherd, Paris: UNESCO Publishing.
Goodman, L.A. (1961) 'Snowball sampling', *The Annals of Mathematical Statistics*, 32, 1: 148–70.
Gordon, T.J. and Helmer, O. (1964) *Report on a Long-Range Forecasting Study*, Santa Monica, CA: The Rand Corporation.
Greimas, A.J. (1987) *On Meaning: Selected Writings in Semiotic Theory*, Theory and

History of Literature, Vol. 38, trans. P.J. Perron and F.H. Collins, Minneapolis, MN: University of Minnesota Press.

Grismer, M.E., Chang, A.C., Grattan, S., Jenkins, D., Phene, C., Rhoades, J.D., Schroeder, E.D., Schwankl, L., Wallender, W.W., Tanji, K.K., Campos, C., Hukkinen, J. and Woodring, C. (1988) *San Joaquin Valley Agriculture and River Water Quality*, No. 3 in a Series on Drainage, Salinity and Toxic Constituents by University of California Committee of Consultants on San Joaquin River Water Quality Objectives, California: The University of California Salinity/Drainage Task Force and Water Resources Center.

Groombridge, B. (ed.) (1992) *Global Biodiversity: Status of the Earth's Living Resources*, Compiled by World Conservation Monitoring Centre, London: Chapman & Hall.

Haas, P.M. (1990) *Saving the Mediterranean: The Politics of International Environmental Co-operation*, New York: Columbia University Press.

Haas, P.M., Keohane, R.O. and Levy, M.A. (1993) *Institutions for the Earth: Sources of Effective International Environmental Protection*, Cambridge, MA: The MIT Press.

Habermas, J. (1975) *Legitimation Crisis*, trans. T. McCarthy, Boston, MA: Beacon Press.

Haigh, N. (1992) 'The European Community and international environmental policy', in A. Hurrell and B. Kingsbury (eds), *The International Politics of the Environment*, Oxford: Oxford University Press, pp. 228–49.

Haila, Y. and Levins, R. (1992) *Humanity and Nature: Ecology, Science and Society*, London: Pluto Press.

Hall, S. (1989a) 'My turn', *West Valley Journal*, September/October, p. 3.

Hall, S. (1989b) 'My turn', *West Valley Journal*, December, p. 3.

Hanemann, M., Lichtenberg, E., Zilberman, D., Chapman, D., Dixon, L., Ellis, G. and Hukkinen, J. (1987) 'Economic implications of regulating agricultural drainage to the San Joaquin River', Appendix G of California State Water Resources Control Board, *Regulation of Agricultural Drainage to the San Joaquin River*, Sacramento, CA: State Water Resources Control Board.

Hannan, M.T. and Freeman, J.H. (1977) 'The population ecology of organizations', *American Journal of Sociology*, 82, 5: 929–64.

Harary, F., Norman, R.Z. and Cartwright, D. (1965) *Structural Models: An Introduction to the Theory of Directed Graphs*, New York: John Wiley.

Harris, T. and Morris, J. (1985) 'Massive US water projects blamed', *Selenium: Toxic Trace Element Threatens the West: The Bee Uncovers Conspiracy of Silence*, reprint from *The Sacramento Bee*, 8–10 September, pp. 3–5.

Harrison, B.E. (1993) *Going Green: How to Communicate Your Company's Environmental Commitment*, Homewood, IL: Business One Irwin.

Harrop, J. (1989) *The Political Economy of Integration in the European Community*, Hampshire: Edward Elgar.

Hart, J. (ed.) (1984) *The New Book of California Tomorrow: Reflections and Projections from the Golden State*, Los Altos, CA: William Kaufmann.

Hart, S.L., Enk, G.A. and Hornick, W.F. (1984), 'Concluding observations and future directions', in S.L. Hart, G.A. Enk and W.F. Hornick (eds) *Improving Impact Assessment: Increasing the Relevance and Utilization of Scientific and Technical Information*, Boulder, CO: Westview Press, pp. 423–34.

Hartshorn, J.K. (1985) 'Down the drain to Kesterson', *Western Water*, March/April, pp. 4–10.

Herman, R., Ardekani, S.A. and Ausubel, J.H. (1989) 'Dematerialization', in J.H. Ausubel and H.E. Sladovich (eds) *Technology and Environment*, Washington, DC: National Academy Press, pp. 50–69.

Hilgard, E.W. (1886) *Alkali Lands, Irrigation and Drainage in their Mutual Relations*, Report by University of California, College of Agriculture, Appendix No. VII to the report for the year 1886, Sacramento, CA: State Office.

Horne, A.J. (1988) 'A research scientist's perspective on the management of Kesterson reservoir: a marsh contaminated with selenium-rich agricultural drain water', *Lake and Reservoir Management*, 4, 2: 187–98.

Hosking, D.-M. and Morley, I.E. (1991) *A Social Psychology of Organizing: People, Processes and Contexts*, New York: Harvester Wheatsheaf.

Hukkinen, J. (1990) *Unplugging Drainage: Toward Sociotechnical Redesign of San Joaquin Valley's Agricultural Drainage Management*, dissertation submitted for Ph.D. in Civil Engineering, Berkeley, CA: University of California.

Hukkinen, J. (1991a) 'Irrigation-induced water quality problems: can present agencies cope?', *Journal of Soil and Water Conservation*, 46, 4: 276–8.

Hukkinen, J. (1991b) *Review of Models Linking Agricultural Drainwater Quality to Management Practices*, report to US Environmental Protection Agency, Berkeley, CA: University of California, Department of Agricultural and Resource Economics.

Hukkinen, J. (1991c) 'Sociotechnical analysis of irrigation drainage in central California', *Journal of Water Resources Planning and Management*, 117, 2: 217–34.

Hukkinen, J. (1993a) 'Bayesian analysis of agricultural drainage problems in California's San Joaquin Valley', *Journal of Environmental Management*, 37: 183–200.

Hukkinen, J. (1993b) 'Institutional distortion of drainage modeling in Arkansas River Basin', *Journal of Irrigation and Drainage Engineering*, 119, 5: 743–55.

Hukkinen, J. (1995a) 'Corporatism as an impediment to ecological sustenance: the case of Finnish waste management', *Ecological Economics*, 15, 1: 59–75.

Hukkinen, J. (1995b) 'Green virus: exploring the environmental product concept', *Business Strategy and the Environment*, 4, 3: 135–44.

Hukkinen, J. (1995c) 'Long-term environmental policy under corporatist institutions', *European Environment*, 5, 4: 98–105.

Hukkinen, J., Roe, E. and Rochlin, G.I. (1988) *When Water Doesn't Mean Power: How Government Can Better Handle Uncertainty and Polarization Related to Agricultural Drainage in the San Joaquin Valley*, Berkeley, CA: University of California.

Hukkinen, J., Roe, E. and Rochlin, G.I. (1990) 'A salt on the land: a narrative analysis of the controversy over irrigation-related salinity and toxicity in California's San Joaquin Valley', *Policy Sciences*, 23, 4: 307–29.

Hume, D. (1978) [1739–40] *A Treatise of Human Nature*, 2nd ed. [1888], ed. L.A. Selby-Bigge and P.H. Nidditch, Oxford: Oxford University Press.

Hurrell, A. and Kingsbury, B. (1992) 'The international politics of the environment: an introduction', in A. Hurrell and B. Kingsbury (eds), *The International Politics of the Environment*, Oxford: Oxford University Press, pp. 1–47.

Imhoff, E. (1989) 'Preface' of San Joaquin Valley Drainage Program, *Preliminary*

Planning Alternatives for Solving Agricultural Drainage and Drainage-Related Problems in the San Joaquin Valley, Sacramento, CA: San Joaquin Valley Drainage Program, pp. iii–iv.

Jantsch, E. (1967) *Technological Forecasting in Perspective*, Paris: Organization for Economic Co-operation and Development (OECD).

Jantsch, E. (1985) *The Self-Organizing Universe: Scientific and Human Implications of the Emerging Paradigm of Evolution*, Oxford: Pergamon Press.

Jenkins, D. (1986) 'Treatment technology relative to San Joaquin River water quality objectives', Appendix H of California State Water Resources Control Board (1987) *Regulation of Agricultural Drainage to the San Joaquin River*, Sacramento, CA: State Water Resources Control Board.

Johns, G.E. and Watkins, D.A. (1989) 'Regulation of agricultural drainage to San Joaquin River', *Journal of Irrigation and Drainage Engineering*, 115, 1: 29–41.

Kahrl, W.L. (ed.) (1979) *The California Water Atlas*, Sacramento, CA: State of California.

Kelley, R.L. and Nye, R.L. (1984) 'Historical perspective on salinity and drainage problems in California', *California Agriculture*, 38, 10: 4–6.

Kim, J.H. and Pearl, J. (1983) 'A computational model for causal and diagnostic reasoning in inference systems', in *Proceedings of the Eighth International Joint Conference on Artificial Intelligence*, Karlsruhe, West Germany, 8–12 August, Los Altos, CA: William Kaufmann, pp. 190–3.

Koivukoski, E. (1992) *10 Years of Hazardous Waste Management* (Ongelmajätehuoltoa 10 vuotta), in Finnish, Riihimäki: Mainos Wallin.

Konikow, L.F. and Bredehoeft, J.D. (1974) 'Modeling flow and chemical quality changes in an irrigated stream–aquifer system', *Water Resources Research*, 10, 3: 546–62.

Konikow, L.F. and Person M. (1985) 'Assessment of long-term salinity changes in an irrigated stream–aquifer system', *Water Resources Research*, 21, 11: 1611–24.

Kovda, V.A. (1983) 'Loss of productive land due to salinization', *Ambio*, 12, 2: 91–3.

Krauskopf, K.B. (1990) 'Disposal of high-level nuclear waste: is it possible?' *Science*, 249: 1231–2.

Kristof, N.D. and WuDunn, S. (1994) *China Wakes: The Struggle for the Soul of a Rising Power*, New York: Times Books.

Landesmann, M. and Vartiainen, J. (1992) 'Social corporatism and long-term economic performance', in J. Pekkarinen, M. Pohjola and B. Rowthorn (eds) *Social Corporatism: A Superior Economic System?* Oxford: Clarendon Press, pp. 210–41.

LaPorte, T.R. (1987) *High Reliability Organizations: The Dimensions of the Research Challenge*, Berkeley, CA: University of California, Institute of Governmental Studies.

Lee, E.W., Nishimura, G.H. and Hansen, H.L. (1988a), *Technical Report: Agricultural Drainage Water Treatment, Reuse, and Disposal in the San Joaquin Valley*, Part I, *Treatment Technology*, Sacramento, CA: San Joaquin Valley Drainage Program.

Lee, E.W., Nishimura, G.H. and Hansen, H.L. (1988b) *Technical Report: Agricultural Drainage Water Treatment, Reuse, and Disposal in the San Joaquin Valley*, Part II, *Reuse and Disposal*, Sacramento, CA: San Joaquin Valley Drainage Program.

Lefkoff, L.J. and Gorelick, S.M. (1990a) 'Simulating physical processes and economic behavior in saline, irrigated agriculture: model development', *Water Resources Research*, 26, 7: 1359–69.

Lefkoff, L.J. and Gorelick, S.M. (1990b) 'Benefits of an irrigation water rental market in a saline stream–aquifer system', *Water Resources Research*, 26, 7: 1371–81.

Letey, J., Roberts, C., Penberth, M. and Vasek, C. (1986) *An Agricultural Dilemma: Drainage Water and Toxics Disposal in the San Joaquin Valley*, Oakland, CA: University of California, Division of Agriculture and Natural Resources.

Levy, M. (1996) 'Assessing the effectiveness of international environmental institutions', *Global Environmental Change*, 6, 4: 395–7.

Liebert, H. (1988) 'Cleanup halted at Kesterson', *San Francisco Chronicle*, April 1.

Lilius, A.-L. (1992) 'Only business in mind' ('Vain bisnes mielessä'), in Finnish, *Talouselämä*, No. 25, 28 August, p. 27.

Liu Yaoqi (1994) 'Analysis of present state and control strategies of China's industrial pollution', in Mandarin, *Environmental Protection of Shanghai*, No. 2.

McCarthy, J. (1985) 'Epistemological problems of artificial intelligence', in R.J. Brachman and H.J. Levesque (eds), *Readings in Knowledge Representation*, Los Altos, CA: Morgan Kaufmann, pp. 23–30. (First appeared in Proceedings of the 5th International Joint Conference on Artificial Intelligence, Cambridge, MA, 1977, pp. 1038–44.)

McCarthy, T. (1975) 'Translator's introduction', in J. Habermas, *Legitimation Crisis*, Boston, MA: Beacon Press, pp. vii–xxiv.

McKeown, K.R. (1985) *Text Generation: Using Discourse Strategies and Focus Constraints to Generate Natural Language Text*, Cambridge: Cambridge University Press.

Mann, M. (1993) 'Nation-states in Europe and other continents: diversifying, developing, not dying', *Daedalus*, 122, 3: 115–40.

Mannermaa, M. (1991) *Evolutionary Futures Research: A Study of Paradigms and their Methodological Characteristics in Futures Research* (Evolutionaarinen tulevaisuudentutkimus: tulevaisuudentutkimuksen paradigmojen ja niiden metodologisten ominaisuuksien tarkastelua), in Finnish with English summary, Acta Futura Fennica, no. 2, Helsinki: Finnish Society for Futures Studies.

March, J.G. and Simon, H.A. (1994) *Organizations*, 2nd ed., Cambridge, MA: Blackwell.

Mazmanian, D. and Morell, D. (1988) 'The elusive pursuit of toxics management', *The Public Interest*, winter, No. 90, 81–98.

Ministry of Housing, Physical Planning and the Environment (1989) *To Choose or to Lose: National Environmental Policy Plan*, The Hague: SDU Publishers.

Ministry of Housing, Spatial Planning and the Environment (1994) *The Netherlands' National Environmental Policy Plan 2*, VROM 93561/b/4–94, The Hague: Ministry of Housing, Spatial Planning and the Environment.

Ministry of the Interior (1983) *Directives on Planning, Permitting, and Disclosure Procedures in Waste Management* (Jätehuollon suunnitelma-, lupa- ja ilmoitusmenettelyjä koskevat ohjeet), in Finnish, Publication B: 6, Helsinki: Ministry of the Interior of Finland, Department of Environmental Protection.

Modern Chinese–English Dictionary (1988) Beijing: Foreign Language Teaching and Research Press.

Moore, J.F. (1993) 'Predators and prey: a new ecology of competition', *Harvard Business Review*, May–June, pp. 75–86.

Moore, S.B. (1989) 'Selenium in agricultural drainage: essential nutrient or toxic threat?' *Journal of Irrigation and Drainage Engineering*, 115, 1: 21–8.

Morgan, G. (1986) *Images of Organization*, Beverly Hills: Sage.

MWD Focus (1989) 'MWD and IID agree on landmark water swap', No. 1, p. 1.

Nadaï, A. (1994) 'The greening of the EC agrochemical market: regulation and competition', *Business Strategy and the Environment*, 3, 2: 34–42.

National Environmental Protection Agency (1994a) 'China State of the Environment Bulletin 1993' (in Mandarin), *Environmental Protection Research*, No. 4.

National Environmental Protection Agency (1994b) 'Outline of national environmental protection work 1993–1998' (Guo Jia Huan Jing Bao Hu Gong Zuo Gang Yao 1993–1998), *Environmental Protection* (Huan Jing Bao Hu), No. 3, pp. 6–10.

National Research Council (1989) *Irrigation-Induced Water Quality Problems: What Can Be Learned from the San Joaquin Valley Experience*, Washington, DC: National Academy Press.

National Research Council (1990) *Ground Water Models: Scientific and Regulatory Applications*, Washington, DC: National Academy Press.

Negoita, C.V. (1985) *Expert Systems and Fuzzy Systems*, Menlo Park, CA: The Benjamin/Cummings Publishing.

Nelkin, D. (1984) [1979] 'Science, technology, and political conflict: analyzing the issues', in D. Nelkin (ed.), *Controversy: Politics of Technical Decisions*, 2nd ed., Beverly Hills, CA: Sage Publications, pp. 9–24.

Nilsson, N.J. (1980) *Principles of Artificial Intelligence*, Palo Alto, CA: Tioga Publishing.

Norgaard, R.B. (1994) *Development Betrayed: The End of Progress and a Coevolutionary Revisioning of the Future*, London: Routledge.

North, D. (1992) *Institutions, Institutional Change and Economic Performance*, Cambridge: Cambridge University Press.

Nousiainen, A. (1992) 'Waste must be reduced' ('Jätettä on vähennettävä'), in Finnish, *Helsingin Sanomat*, 28 March, p. A15.

Nugent, N. (1989) *The Government and Politics of the European Community*, Durham, NC: Duke University Press.

Office of Technology Assessment (1989) *Facing America's Trash: What Next for Municipal Solid Waste?* Washington, DC: Congress of the United States.

Ohlendorf, H.M. (1986) 'Aquatic birds and selenium in the San Joaquin Valley', *Selenium and Agricultural Drainage: Implications for San Francisco Bay and the California Environment*, Proceedings of the Second Selenium Symposium, Berkeley, CA, 23 March 1985, Tiburon, CA: The Bay Institute of San Francisco, pp. 14–24.

Opschoor, H. and van der Straaten, J. (1993), 'Sustainable development: an institutional approach', *Ecological Economics*, 7, 3: 203–22.

Osborn, A.F. (1963) *Applied Imagination: Principles and Procedures of Creative Problem-Solving*, 3rd revised ed., New York: Charles Scribner's Sons.

Oster, J., Fulton, A., Goldhamer, D.A., Grimes, D.W., Hanson, B.R., Hoffman, G., Knapp, K., Letey, J., Weir, B., Tanji, K.K., Campos, C., Woodring, C., Wallender, W.W., Wooley, D., O'Neill, T., Stone, B. and Jones, R. (1988) *Opportunities for Drainage Water Reduction*, No. 1 in a series on drainage, salinity and toxic constituents by University of California Committee of Consultants on Drainage Water Reduction. California: The University of California Salinity/ Drainage Task Force and Water Resources Center.

Östlund, S. (1994) 'The limits and possibilities in designing the environmentally sustainable firm', *Business Strategy and the Environment*, 3, 2: 21–33.

Ostrom, E. (1994) *Governing the Commons: The Evolution of Institutions for Collective Action*, Cambridge: Cambridge University Press.

Ostrom, E. (1995) 'Designing complexity to govern complexity', in S. Hanna and M. Munasinghe (eds) *Property Rights and the Environment: Social and Ecological Issues*, Washington, DC: The World Bank, pp. 33–45.

Oswald, W.J. (1991) 'Terrestrial approaches to integration of waste treatment', *Waste Management and Research*, 9: 477–84.

Palokangas, R., Tarukannel, V. and Nuuja, I. (1993) *The New Administration and Legislation of Environmental Protection* (Uusi ympäristönsuojelun hallinto ja lainsäädäntö), in Finnish, Jyväskylä: Ympäristö-Tieto Ky.

Pantzar, M. (1992) 'The growth of product variety – a myth?', *Journal of Consumer Studies and Home Economics*, 16: 345–62.

Pearce, D.W. and Turner, R.K. (1990) *Economics of Natural Resources and the Environment*, Baltimore, MD: The Johns Hopkins University Press.

Pearl, J. (1986) 'Fusion, propagation, and structuring in belief networks', *Artificial Intelligence*, 29, 3: 241–88.

Pearl, J. (1988) *Probabilistic Reasoning in Intelligent Systems: Networks of Plausible Inference*, San Mateo, CA: Morgan Kaufmann.

Pekkarinen, J., Pohjola, M. and Rowthorn, B. (1992) 'Social corporatism and economic performance: introduction and conclusions', in J. Pekkarinen, M. Pohjola and B. Rowthorn (eds) *Social Corporatism: A Superior Economic System?*, Oxford: Clarendon Press, pp. 1–23.

Perrow, C. (1970) *Organizational Analysis: A Sociological View*, Belmont, CA: Wadsworth Publishing Company.

Perrow, C. (1984) *Normal Accidents: Living with High-Risk Technologies*, New York: Basic Books.

Peterson, W.C. (1973) *Elements of Economics*, New York: W.W. Norton.

Pohjanpalo, O. (1991) 'Waste incinerator knocked out' ('Jätteenpolttolaitos tyrmättiin'), *Helsingin Sanomat*, 20 December, p. B1.

Popper, K.R. (1977) [1966] *The Open Society and Its Enemies*, Vol. I, *The Spell of Plato*, 5th ed., London: Routledge & Kegan Paul.

Powledge, F. (1983) *Water: The Nature, Uses, and Future of Our Most Precious and Abused Resource*, New York: Farrar Straus Giroux.

Press, S.J. (1989) *Bayesian Statistics: Principles, Models, and Applications*, New York: John Wiley & Sons.

Radosevich, G.E., Nobe, K.C., Allardice, D. and Kirkwood C. (1976) *Evolution and Administration of Colorado Water Law: 1876–1976*, Fort Collins, CO: Water Resources Publications.

Rawls, J. (1971) *A Theory of Justice*, Cambridge, MA: The Belknap Press of Harvard University Press.

Redclift, M. (1992) *Sustainable Development: Exploring the Contradictions*, London: Routledge.

Reed, D. (1996) *Structural Adjustment, the Environment, and Sustainable Development: Executive Summary*, Washington, DC: WWF-International.
Reisner, M. (1987) *Cadillac Desert: The American West and its Disappearing Water*, New York: Penguin Books.
Reisner, M. (1989) 'The emerald desert', *Greenpeace*, 14, 4: 6–10.
Riffaterre, M. (1983) *Text Production*, trans. T. Lyons, New York: Columbia University Press.
Robinson, N.A. (ed.) (1993) *Agenda 21: Earth's Action Plan*, IUCN Environmental Policy & Law Paper No. 27, New York: Oceana.
Robinson, D.F. and Foulds, L.R. (1980) *Digraphs: Theory and Techniques*, New York: Gordon and Breach Science Publishers.
Rochlin, G.I. (1990) 'High-tech weapons systems are two-edged swords', *Public Affairs Report*, 31, 6: 1, 10–11.
Rochlin, G.I., LaPorte, T.R. and Roberts, K.H., (1987) 'The self-designing high-reliability organization: aircraft carrier flight operations at sea', *Naval War College Review*, Autumn.
Roe, E.M. (1991) 'Overseas perspectives for managing irrigation drainage in California', *Journal of Irrigation and Drainage Engineering*, 117, 3: 350–60.
Roe, E. (1994) *Narrative Policy Analysis: Theory and Practice*, Durham, NC: Duke University Press.
Roe, E. (1996) 'Sustainable development and Girardian economics', *Ecological Economics*, 16, 2: 87–93.
Rohwer, J. (1992) 'When China wakes', *The Economist: A Survey of China*, 28 November, pp. 1–22.
Rudischhauser, K. (1992) 'Developing a common waste management policy in the European Community', paper for the *IUAPPA 9th World Clean Air Congress and Exhibition*, Montreal, Canada, 30 August–4 September, Brussels: Commission of the European Communities.
Rudolph, E. (1988) 'Connective relations – connective expressions – connective structures', in J. Petöfi (ed.), *Text and Discourse Constitution: Empirical Aspects, Theoretical Approaches*, Berlin: Walter de Gruyter, pp. 97–133.
Rutherford, M. (1996) *Institutions in Economics: The Old and the New Institutionalism*, Cambridge: Cambridge University Press.
RVFs Framtidskommitté (1991) *Swedish Waste Management, a Futures Study in Scenario Form* (Svensk avfallshantering, en framtidsstudie i scenarioform), in Swedish, Malmö: Svenska Renhållningsverks-Föreningen/RVF.
Ryan, M. and Flavin, C. (1995) 'Facing China's limits', in L.R. Brown *et al.*, *State of the World 1995*, New York: W.W. Norton, pp. 113–31.
San Francisco Chronicle (1989) 'Backing for big valley drain', 22 September, p. A4.
San Joaquin Valley Drainage Advisory Group (1969) *Final Report*, Fresno, CA: California Department of Water Resources.
San Joaquin Valley Drainage Program (1987) *Developing Options: An Overview of Efforts to Solve Agricultural Drainage and Drainage-Related Problems in the San Joaquin Valley*, Sacramento, CA: San Joaquin Valley Drainage Program.
San Joaquin Valley Drainage Program (1989) *Preliminary Planning Alternatives for Solving Agricultural Drainage and Drainage-Related Problems in the San Joaquin Valley*, Sacramento, CA: San Joaquin Valley Drainage Program.
San Joaquin Valley Drainage Program (1990) *A Management Plan for Agricultural*

Subsurface Drainage and Related Problems on the Westside San Joaquin Valley, Sacramento, CA: US Department of the Interior and California Resources Agency.

San Joaquin Valley Drainage Program Status Report (1987) 'Kesterson Program', April, No. 7, pp. 4–5.

San Joaquin Valley Drainage Program Status Report (1988a) 'Kesterson Program', February, No. 10, p. 3.

San Joaquin Valley Drainage Program Status Report (1988b) 'Kesterson Program', September, No. 11, p. 3.

San Joaquin Valley Drainage Program Status Report (1989) 'Kesterson Program', January, No. 12, p. 3.

San Joaquin Valley Interagency Drainage Program (1979) *Agricultural Drainage and Salt Management in the San Joaquin Valley, Final Report*, Fresno, CA: Interagency Drainage Program.

Saussure, F. de (1966) [1959] *Course in General Linguistics*, eds C. Bally, A. Sechehaye and A. Riedlinger, trans. W. Baskin, New York: McGraw-Hill.

Schmidheiny, S. (1992) *Changing Course: A Global Business Perspective on Development and the Environment*, Cambridge, MA: The MIT Press.

Schmitt, S.A. (1969) *Measuring Uncertainty: An Elementary Introduction to Bayesian Statistics*, Reading, MA: Addison-Wesley.

Schroeder, R.A., Palawski, D.U. and Skorupa, J.P. (1988) *Reconnaissance Investigation of Water Quality, Bottom Sediment, and Biota Associated with Irrigation Drainage in the Tulare Lake Bed Area, Southern San Joaquin Valley, California, 1986–87*, US Geological Survey Water-Resources Investigations Report 88–4001, Denver, CO: US Geological Survey.

Scott, W.R. (1987) *Organizations: Rational, Natural, and Open Systems*, 2nd ed., Englewood Cliffs, NJ: Prentice-Hall.

Selznick, P. (1984) [1949] *TVA and the Grass Roots: A Study of Politics and Organization*, Berkeley, CA: University of California Press.

Simon, H.A. (1964) [1957] *Administrative Behavior: A Study of Decision-Making Processes in Administrative Organization*, 2nd ed., New York: The Macmillan Company.

Smil, V. (1993) *China's Environmental Crisis: An Inquiry into the Limits of National Development*, Armonk, NY: East Gate.

Soil Conservation Service (1990) *Patterson Hollow USDA Water Quality Hydrologic Unit Area Proposal*, Otero (089) and Pueblo (101) counties, Colorado (08), USGS hydrologic unit 11020005, Junc, United States of America: Soil Conservation Service.

Sutherland, L.P. and Knapp, J.A. (1988) 'The impacts of limited water: a Colorado case study', *Journal of Soil and Water Conservation*, 43, 4: 294–8.

Szalay, L.B., Strohl, J.B., Liu Fu and Pen-Shui Lao (1994) *American and Chinese Perceptions and Belief Systems: A People's Republic of China – Taiwanese Comparison*, New York: Plenum Press.

Tanji, K., Läuchli, A. and Meyer, J. (1986) 'Selenium in the San Joaquin Valley', *Environment*, 28, 6: 6–11, 34–9.

Tarr, D. (1988) 'The Journey of a million acres of cotton may begin with a single drop', *Aqueduct*, 53, 1: 21–5.

Taylor, K.C., Mayewski, P.A., Alley, R.B., Brook, E.J., Gow, A.J., Grootes, P.M.,

Meese, D.A., Saltzman, E.S., Severinghaus, J.P., Twickler, M.S., White, J.W.C., Whitlow, S. and Zielinski, G.A. (1997) 'The Holocene–Younger Dryas transition recorded at Summit, Greenland', *Science*, 278: 825–7.

Taylor, P.S. (1979) *Essays on Land, Water and the Law in California*, New York: Arno Press.

Taylor, S. (1984) *Making Bureaucracies Think: The Environmental Impact Statement Strategy of Administrative Reform*, Stanford, CA: Stanford University Press.

Therborn, G. (1992) 'Lessons from "corporatist" theorizations', in J. Pekkarinen, M. Pohjola and B. Rowthorn (eds) *Social Corporatism: A Superior Economic System?*, Oxford: Clarendon Press, pp. 24–43.

Thompson, J.D. (1967) *Organizations in Action: Social Science Bases of Administrative Theory*, New York: McGraw-Hill.

Thurm, S. (1993a) 'Poison in the ponds', *San Jose Mercury News*, 4 July, pp. 1A, 28A–29A.

Thurm, S. (1993b) 'No easy answers', *San Jose Mercury News*, 6 July, pp. 1E–2E.

Tibbs, H.B.C. (1992) 'Industrial ecology: an environmental agenda for industry', *Whole Earth Review*, winter, pp. 4–19.

Trompeter, K.M. and Suemoto, S.H. (1984) 'Desalting by reverse osmosis at Yuma Desalting Plant', in R.H. French (ed.), *Salinity in Watercourses and Reservoirs*, Proceedings of the 1983 International Symposium on State-of-the-Art Control of Salinity, Salt Lake City, Utah, 13–15 July, Boston, MA: Butterworth Publishers, pp. 427–37.

Tversky, A. and Kahneman, D. (1982) 'Causal schemas in judgements under uncertainty', in D. Kahnemann, P. Slovic and A. Tversky (eds), *Judgement under Uncertainty: Heuristics and Biases*, Cambridge: Cambridge University Press, pp. 117–28.

Uphoff, N. (1986) *Improving International Irrigation Management with Farmer Participation: Getting the Process Right*, Studies in Water Policy and Management, No. 11, Boulder, CO: Westview Press.

Uphoff, N., Ramamurthy, P. and Steiner, R. (1988) *Improving Performance of Irrigation Bureaucracies: Suggestions for Systematic Analysis and Agency Reorientation*, Water Management Synthesis II Project, Ithaca, NY: Cornell University, Irrigation Studies Group.

US Bureau of Reclamation (1945) *Comprehensive Plan for Water Resources Development: Central Valley Basin, California*, Project Planning Report No. 2–4.0–3, November, Washington, DC: US Bureau of Reclamation.

US Bureau of Reclamation (1964) *Alternative Solutions for Drainage: San Luis Unit, Central Valley Project, California*, February, United States of America: US Bureau of Reclamation.

US Committee on Irrigation and Drainage (1987), *How Can Irrigated Agriculture Exist with Toxic Waste Regulations?*, Report of the Panel of Experts, San Diego, CA, 29–30 October 1987, Denver, CO: US Committee on Irrigation and Drainage.

US Department of the Interior (1993) *Audit Report: Department of the Interior Irrigation Drainage Programs*, Report No. 93–I-1302, July, Washington, DC: US Department of the Interior, Office of Inspector General.

US Geological Survey (1990) *Evaluation of Water Quality in the Arkansas River*

Basin in Colorado, revised work plan, 12 June, Pueblo, CO: US Geological Survey.

Uusitalo, L. (1991) *Finns and the Environment: A Study of the Rationality of Economic Behaviour* (Suomalaiset ja ympäristö: Tutkimus taloudellisen käyttäytymisen rationaalisuudesta), in Finnish, Acta Academiae Oeconomicae Helsingiensis, A: 49, Helsinki: The Helsinki School of Economics and Business Administration.

Vahvelainen, S. and Isaksson, K.-E. (1992) *Industrial Waste* (Teollisuuden jätteet), in Finnish with English summary, Helsinki: Statistics Finland.

Vickers, G. (1983) *Human Systems are Different*, London: Harper & Row.

Waste Management Advisory Board (1991) *Development Programme on Municipal Waste Management 2000* (Yhdyskuntien jätehuollon kehittämisohjelma 2000), in Finnish with English abstract, Report 104/1991, Helsinki: Ministry of the Environment of Finland, Environmental Protection Department.

Waste Management Council (1992) *Toward a Ten-Year Programme on Waste Management*, Utrecht: Waste Management Council of the Netherlands.

Waste Management Law Committee Report I (1992) *Proposal for Waste Management Law* (Ehdotus jätelaiksi), in Finnish, Committee Reports 19, Helsinki: Ministry of the Environment of Finland.

Weick, K.E. (1969) *The Social Psychology of Organizing*, Reading, MA: Addison-Wesley.

Weiss, S.M. and Kulikowski, C.A. (1984) *A Practical Guide to Designing Expert Systems*, Totowa, NJ: Rowman & Allanheld.

Welford, R. (1993) 'Breaking the link between quality and the environment: auditing for sustainability and life cycle assessment', *Business Strategy and the Environment*, 2, 4: 25–33.

Welford, R. and Gouldson, A. (1993) *Environmental Management and Business Strategy*, London: Pitman Publishing.

West Valley Journal (1988a) 'Selenium removal project hits snag', February, p. 1.

West Valley Journal (1988b) 'WWD okays $187,000 cogeneration study', June, p. 2.

West Valley Journal (1988c) 'Construction of Harza Facility slated to begin', June, p. 4.

Westcot, D., Rosenbaum, S., Grewell, B. and Belden, K. (1988) *Water and Sediment Quality in Evaporation Basins Used for the Disposal of Agricultural Subsurface Drainage Water in the San Joaquin Valley, California*, Sacramento, CA: Central Valley Regional Water Quality Control Board.

Westlands Water District Drainage Update (1986) November, No. 9.

Westlands Water District Drainage Update (1988a) 'District suspends work on selenium removal project', May, pp. 1–2.

Westlands Water District Drainage Update (1988b) 'Cogeneration/selenium removal process study begins', August, pp. 2–3.

Wilson, J.Q. (1989) *Bureaucracy: What Government Agencies Do and Why They Do It*, USA [n.p.]: Basic Books.

Wilson, J.Q. and Rachal, P. (1977) 'Can the government regulate itself?', *The Public Interest*, No. 46, winter, pp. 3–14.

Wilson, R.J. (1974) *Introduction to Graph Theory*, New York: Academic Press.

Wittfogel, K.A. (1957) *Oriental Despotism: A Comparative Study of Total Power*, New Haven, CT: Yale University Press.

Wolf, S.A. and Allen, T.F.H. (1995) 'Recasting alternative agriculture as a management model: the value of adept scaling', *Ecological Economics* 12, 1: 5–12.

World Commission on Environment and Development (1990) *Our Common Future*, Oxford: Oxford University Press.

World Development Report 1992: Development and the Environment (1992) New York: Oxford University Press.

Worster, D. (1985) *Rivers of Empire: Water, Aridity, and the Growth of the American West*, New York: Pantheon Books.

Wright, D. (1991) 'EC environmental policy: coping with interdependency', *Futures*, 23, 7: 709–23.

Yang Zhou Ji (1994) 'Clean production' (Xing Jie Sheng Chan), *China Environmental Management* (Zhong Guo Huan Jing Guan Li), No. 3, pp. 4–7.

Young, O.R. (1982) *Resource Regimes: Natural Resources and Social Institutions*, Berkeley: University of California Press.

Young, O.R. (1994) *International Governance: Protecting the Environment in a Stateless Society*, Ithaca: Cornell University Press.

INDEX